THE
CARBON
COLLISION
COURSE

Australia's Emissions and Energy Policy Crisis

ANDREW PERRY

Published in Australia by Pursuit Energy and Project Consulting

Printed in Australia

First Edition

National Library of Australia Cataloguing-in-Publication
entry available for this title at nla.gov.au

ISBN: 978-0-9876358-0-8

Cover & Interior Design: Swish Design

Cover Photo: shutterstock.com

Editing & Proofreading: Bill Harper

Reviewer: Donald McMillan FIEAust CPEng; APEC; IntPE(Aus)

The real inconvenient truth is everybody thinks global warming and carbon emissions are serious concerns that somebody else needs to do something about.

Port Talbot Steel Works, Wales, UK.

Closed down in 2015, reducing the UK's reported emissions by 2.3 million tonnes of carbon dioxide equivalent.

Effect on global CO_2 emissions: nil.

Image: shutterstock.com

FOREWORD

Hardly a day goes by without some media coverage on the subject of climate change. Whether it is global warming, extreme weather events, sea level change, or risk of damage to the Great Barrier Reef, climate variations purportedly caused by human-induced greenhouse gas emissions are often cited as the greatest challenge we face this century.

Everybody seems to have an opinion on what should be done about our emissions. However, I wonder if most of us have an appreciation of how dependent on fossil fuels we have become, and what radical changes to our consumption habits need to be undertaken if we are to seriously alter our current emissions trajectory.

This book shows how our everyday consumption of fossil fuel energy products, such as electricity, heating, transportation fuels (petrol and diesel), and airline travel, link to the resulting emissions. It raises the questions of whether renewable energy is realistically capable of fulfilling all our future energy needs, and at what cost? It also looks at whether cutting emissions from heavy industrial activity in some jurisdictions is simply a case of 'squeezing the balloon' (where the emission is simply outsourced to jurisdictions with less onerous reduction targets).

It is essential reading for anybody interested in the emissions policy debate, one that has caused considerable political turmoil over the past 10 years in Australia. It is highly recommended to anyone who wants to understand what political parties' and lobby groups' policy proposals on emissions really mean.

Donald McMillan, FIEAust CPEng; APEC; IntPE(Aus)

AIM OF THIS BOOK

It was assumed in the *Independent Review into the Future Security of the National Electricity Market: Blueprint for the Future, Commonwealth of Australia 2017*[1] that the Australian Government had a policy to commit to the so-called Paris Agreement emission reduction target: "Australia has committed to reduce its emissions by 28 per cent below our 2005 levels by 2030, and ongoing reductions towards zero emissions in the second half of the century".

At the time of the review, no policy or legislation was in place confirming any commitment by Australia to meet the Paris emission reduction target. An emission reduction policy was in the Turnbull Government's proposed National Energy Guarantee (NEG), which was intended to reduce Australia's emissions towards the Paris target. However, factions in the Liberal National Party were strongly opposed to any commitment to further emission reduction targets. This disagreement seemed to be the primary cause of the 2018 LNP leadership crisis, and the subsequent removal of Malcolm Turnbull as Prime Minister of Australia.

This book will help you understand the basics of how carbon emissions work, and their link to energy consumption and energy policy. It provides detailed insights into these areas of the emissions policy debate:

1. Has Australia done too much, the right amount, or too little to reduce greenhouse gas emissions over the past 28 years of the Kyoto Protocol?

2. Is renewable energy really cheaper than fossil fuel?

3. Should Australia be transitioning towards a zero-emission economy by the second half of this century?

4. Is it even possible for Australia to transition towards a zero-emission economy by the second half of this century?

5. Should Australia increase the price of petrol and diesel by around 70 c/L by imposing higher duties to the same level of taxation as in the UK and Western Europe?

6. Would such a proposal get through an election?

7. Aluminium smelting accounts for around 7% of Australia's emissions. Should aluminium production in Australia stop, either through a carbon reduction mandate or a carbon tax?

8. Should Australia stop exporting high direct carbon emission intense materials such as coal and liquefied natural gas (LNG)?

9. Should Australia stop exporting high indirect carbon emission intense materials such as iron ore and bauxite/aluminium ores?

10. Australia's contribution to worldwide greenhouse gas emissions is about 1.2%. Would any action taken by Australia to reduce CO_2e emissions have any significant impact on global emissions, either directly or by influence?

11. Some administrations (most notably the USA) have come to the conclusion that emission reductions can only be achieved at a cost to their economy they aren't willing to bear, and have pulled out of or are revising their emission reduction commitments. Should Australia follow suit?

Like most political issues, the answers to these questions aren't exactly black and white. It's up to the reader to form their own views after considering the all information available. But one thing seems to be quite clear: there is no silver bullet solution to the issue of emissions reduction.

Disclaimer

This book isn't about whether anthropogenic (human-induced) carbon dioxide emissions have affected the climate beyond the natural Earth and solar system cycles that have changed the Earth's climate over the history of the planet. It's about quantifying our carbon emissions and the challenges of reducing them.

It has been published without sponsorship or funding from any lobby or industry association groups. And it's neither for nor against renewable or fossil fuel energy. It simply provides an impartial analysis of the challenges faced in understanding our energy consumption habits and their resulting emissions.

The book collates data and analysis from numerous sources to give a high-level overview of the emissions and energy debate. Due to availability, some data sources are not perfectly aligned. However, the errors due to this are negligible. For example, the analysis of Australian and UK emission and energy consumption profiles are not exactly aligned in terms of timing. But the inaccuracies due to this are negligible compared the absolute differences in the profiles, and it does not affect the conclusions of the analysis.

The data in the book is presented as as accurately as possible at the time of preparation (2018). Data in the emissions and energy sector changes rapidly, and no responsibility can be taken for any inaccuracies in any of the data presented.

CONTENTS

GLOSSARY OF TERMS

t	tonne (1,000 kg)
Mt	million tonnes
Bt	billion tonnes (1,000 million tonnes)
Tt	trillion tonnes (one million million tonnes)
CO_2	Carbon Dioxide, actual Carbon Dioxide emission
CO_2e	Carbon Dioxide equivalent, quantity of greenhouse gas equivalent to CO_2
GHG	Greenhouse Gas
GWP	Global Warming Potential, equivalence of other GHGs as a ratio to that of CO_2
kJ	kilojoule, standard unit of energy
MJ	megajoule, 1,000 kilojoules
GJ	gigajoule, 1,000,000 kilojoules
kW	kilowatt, standard unit of power (=1 kJ/second)
MW	megawatt, 1,000 kilowatts
kWh	kilowatt-hour, standard unit of electrical (or work) energy (=3,600 kJ)
MWh	megawatt-hour, 1,000 kWh
C	Carbon
HC	Hydrocarbon, the main constituent of fossil fuel
LCOE	Levelised Cost of Electricity
RE	Renewable Energy, or renewables

Dispatchable	Available on demand, not subject to ambient conditions
VRE	Variable Renewable Energy, new term for non-dispatchable renewables
Solar PV	Solar photovoltaic, the basic generation technology used in solar panels
LULUCF	Land Use, Land Use Change, and Forestry, emissions from land use changes
CSIRO	Commonwealth Scientific and Industrial Research Organisation
NGGI	National Greenhouse Gas Inventory, Australia GHG Reporting system
IPCC	Intergovernmental Panel on Climate Change, the UN's scientific and reporting organisation for emissions and climate change

- The term "carbon emission" in this book generally refers to carbon dioxide emission (CO_2) or carbon dioxide equivalent (CO_2e).

- The use of the term "per capita" in this book generally means "per capita year" unless otherwise stated.

- Fugitive emissions are escaping methane gas (the main component of natural gas) associated with all fossil fuel extraction and use. Methane gas has a global warming potential (GWP) 25 times that of carbon dioxide (on a 100 year basis)

- The term "The West" refers to the USA and Canada, the EU and the UK, Australia, New Zealand and Japan. The term "Non-West" refers to everywhere else.

INTRODUCTION

Every day the world consumes about 12 million tonnes of oil (around 85 million barrels), 21 million tonnes of coal, and 7.5 million tonnes of natural gas, resulting in about 90 million tonnes of carbon dioxide emissions. That's about 12 $kgCO_2$ per day for every person on the planet.

That might not sound like much, but it's about 12 times the amount of CO_2 we exhale every day simply by breathing (about 1 $kgCO_2$/day). But that's just the world's average. In the West, fossil fuel consumption emits 28 $kgCO_2$ per capita day – 28 times the natural breathing rate. And at 60 $kgCO_2e$ per day (22 tCO_2e per year), Australia has the highest per capita greenhouse gas emission footprint of all Western industrialised nations.

This book explains in simple terms how our everyday consumption links to resulting emissions, and why Australia's emissions are so high compared with many of its Western peers. It also will explain how despite having such high emissions, Australia easily beat its Kyoto Protocol round one emission target without having to take any proactive measures on fossil fuel consumption or efficiency. The Paris emission reduction target is far more onerous, and it will not be possible to meet it without dramatic changes to Australia's energy policy. Lack of a coherent policy on energy and associated emissions was once again a cause of a political crisis in Australia in 2018.

We are constantly reminded about the impact of man-made greenhouse gas emissions on the Earth's environment, particularly the risk of global warming, and the severe consequences it could have for life on Earth. The world's political and scientific communities have been contemplating what needs to be done to mitigate carbon emissions for decades. But so far very little has been achieved, and emissions keep rising year after year.

Despite reductions in some jurisdictions, globally greenhouse gas emissions have risen consistently over the past 65 years by a compound interest rate of around 2.5%. Achieving real emission reductions has been likened to squeezing a balloon: squeeze it at one end and it just bulges out at the other. For example, efficiency improvements in cars and aircraft made over recent years haven't resulted in emission reductions. It has enabled us to travel more, offsetting any gain that might have been achieved.

There's a fierce debate in Australia about whether the country should be transitioning to a "low-emission economy". As already mentioned, Australia has the highest per capita greenhouse gas footprint of all Western industrialised nations. At 22 tCO_2e per capita year, that's about 70% higher than the average of the West. Around 80% of Australia's greenhouse gas footprint is associated directly with fossil fuel consumption – 17 tCO_2 per capita – and has shown no reduction over the past 25 years.

Renewable energy is espoused as the solution to fossil fuel emissions, and certain sectors in the energy debate claim it is already cheaper than fossil fuel. If that's the case, emission reductions should be easy, and renewables should be booming without government subsidy. However, the official studies into the cost of electricity in Australia haven't supported this position. They actually conclude that coal-fired power generation is currently the cheapest form of electricity.

Many sectors of the media and even the scientific community don't report this. Instead they focus on the headline claims. This may be misleading the public and some of our politicians into believing emission reductions will be cheap and easy. Some administrations (most notably the USA) have come to the conclusion that emission reductions can only be achieved at a cost to their economy they aren't willing to bear, and have either pulled out of or refused to take on further reduction commitments.

However high Australia's internal fossil fuel CO_2 emissions may seem, they are paltry compared with the emissions resulting from Australia's fossil fuel exports. Over the past 30 years, Australia has been expanding fossil fuel exports, with current volumes resulting in about 950 $MtCO_2$ being emitted every year. That's around 40 tCO_2 per capita – almost double the internal emission. International reporting convention on greenhouse gases requires jurisdictions to report emissions resulting only from their internal consumption. In other words, Australian fossil fuels exported and burnt in other countries are not included.

But any emission reductions from reduced coal consumption in Australia will do nothing for global emissions if the coal displaced is simply exported to be burned elsewhere. It's another example of squeezing the balloon. Politicians so far have focussed mostly on the push for renewables, but fossil fuel exports are the elephant in the room in the Australian emissions debate.

At about eight tCO_2e per capita, the UK has one of the lowest emission footprints of the Western industrialised nations. Unlike Australia, the UK's per capita CO_2 emissions from fossil fuel consumption have reduced by more than 30% over the Kyoto Protocol target years. Page v of this book shows the Port Talbot Steel Works in Wales, UK. It was closed down in 2015, resulting in the UK's reported emissions reducing by 2.3 $MtCO_2e$.

However, that doesn't mean global emissions reduced by the same amount. The UK's demand for steel didn't change when the facility ceased production. All that happened was the same quantity of steel was imported from major steel exporting countries, the largest being China and Japan. Although the UK has made tangible reductions in fossil fuel emissions, it is estimated that after taking net imports and exports into account, the UK's emissions resulting from general economic activity have actually risen substantially over the Kyoto years[2]. Through global trade, the UK has been able to outsource its emissions from heavy industrial activity, in many cases to countries that have less onerous emission reduction commitments, or none at all.

It is estimated that globally we have emitted 2,000 billion tonnes of carbon dioxide equivalent (BtCO$_2$e) since the start of the modern industrial revolution in 1850. It is also estimated that to stay within an anticipated rise in global temperatures of less than 2°C, we can only emit a further 1,000 BtCO$_2$e beyond our current emissions. The problem we face is the last 1,000 BtCO$_2$e we emitted occurred in only the past 40 years. Even if we can gradually cut our emissions from today's level over time to zero, we'll use up the remaining emission budget in about 50 years (~=1,000/(40/2)). In the same period we emitted the last 1,000 BtCO$_2$e, the world's population grew from 4.2 to 7.4 billion, with most of this growth occurring in the developing world. This book looks at the magnitude of the emissions problem on a global scale, and the challenges we face in reducing consumption to a level that might have any real impact on atmospheric CO$_2$ levels.

There are two carbon collision points looming:

- There are far more fossil fuel reserves available than the remaining CO$_2$ emission budget allows for – around three times as much probable reserves, and up to 10 times as much when possible reserves are included[2].

- Many of the world's new citizen's in the Non-West desire wealth and prosperity that the exploitation of cheap fossil fuel energy has brought the West.

Access to cheap fossil fuel energy has allowed those living in the West to switch on an air conditioner when they feel like it, drive cars when and where they want, and take long-distance air travel holidays when they wish to. Unfortunately, there's no escaping the fact that to cut emissions we need to cut consumption. That is the real inconvenient truth.

The problem is many people think global warming and emissions are serious concerns that somebody else needs to do something about.

For Australia to be a player in the push to a low-emission world economy, its needs to take a position on the future role of fossil fuel and mineral exports in the Australia economy, and how to replace the wealth it generates.

The current emission policy crisis in Australia isn't just about emissions. It's about which direction to take the whole economy in during the next 25 years.

CHAPTER 1

The world in billions – Part 1:
Population and per capita emissions

Anthropogenic (human-induced) greenhouse gas emissions (GHG) are currently (around 2015) estimated to be about 44 billion tonnes of carbon dioxide equivalent (CO_2e) per year. Around 33 billion tonnes (75%) of these emissions are directly attributable to fossil fuel consumption. These numbers can be hard to conceptualise, so let's use per capita figures to put them in an everyday context.

The human population of the world is estimated to be about 7.4 billion people (also around 2015). So our CO_2e emissions are about six tonnes per capita (44/7.4). This is equivalent to about 16 kg of CO_2e emission per capita per day, or 12 kg of CO_2 per capita from fossil fuels. As I said in the introduction, we exhale about one kilogram of CO_2 per day through food energy consumption. So our fossil fuel consumption emission is about 12 times the amount we produce from breathing.

However, that's just the world's average. In the West, fossil fuel consumption results in emission of around 28 kg of CO_2 per capita day – 28 times the natural breathing rate. And around 80% of the West's GHG footprint is associated directly with fossil fuel consumption.

The reason the West consumes so much fossil fuel energy is simple. It has been developed constantly over the past 150 years to be such a convenient way to provide a high rate of useful power. A human labourer can generate about 0.1 kilowatt (kW) of power, though only for around seven hours a day, five days a week. In comparison, a typical car using liquid fossil fuel petrol can operate at cruising speed of 90 kph with a power output of about 20 kW effectively continuously. That's equivalent to roughly 200 human workers, or about 26 horses (one horsepower being equivalent to 0.746 kW). And that's just the power output at cruising speed. During maximum acceleration or a steep climb, the vehicle's power output may reach 100kW.

The main advantage the combination of internal combustion engine and fossil fuel has over human or horsepower is its power-to-weight ratio. An average 75 kg human has a power-to-weight ratio of 1.3 W/kg. A 750 kg car with a 100kW engine has a power-to-weight ratio of 133W/kg – 100 times that of a human.

Having this amount of power available is incredibly convenient. While driving to the shops in a car consumes far more energy than walking, the car will get there in about a tenth of the time. And it can carry a lot more than a person could carry themselves. What's more, the calorific energy from fossil fuel is cheaper to produce than food energy.

High-energy-density fossil fuel is convenient and relatively cheap, making it an attractive (if not addictive) energy option.

Population Growth

The world's population is estimated to be around 7.4 billion.

Having this many human beings on the planet is a recent phenomenon. Here's the estimated growth of human population over the past 12,000 years since the beginning of primitive human agriculture (around the end of the last Great Ice Age).

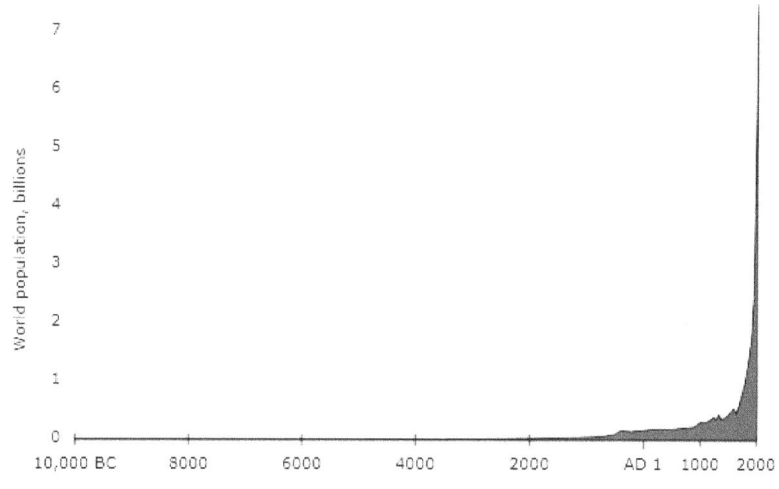

Source: Wikipedia – World Population

The world population didn't reach one billion until around 1800. And in 1950 it was only 2.8 billion.

Now here's a chart comparing this population growth over just the last 200 years, with fossil fuel GHG emissions (expressed as billion tonnes of carbon (1 Bt-carbon is equivalent to 3.67 Bt-carbon dioxide, see chapter 2)) and estimates of the rising atmospheric CO_2 concentration in parts per million (ppm).

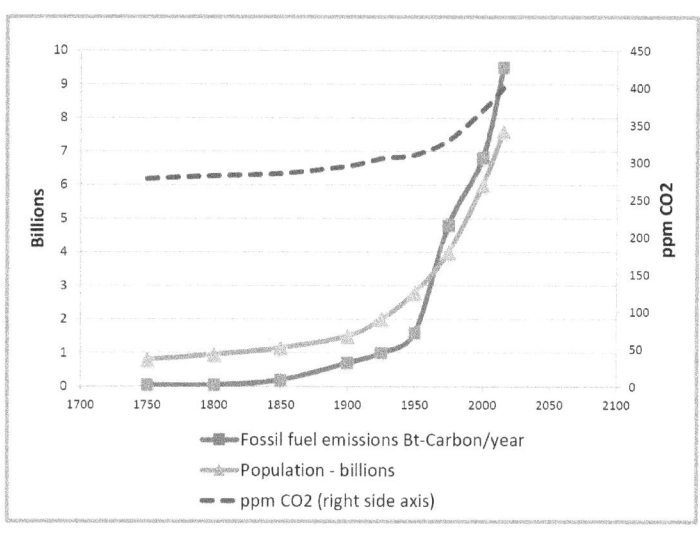

It's easy to see a correlation between population, carbon emissions and atmospheric CO_2 concentration. But this doesn't prove there's a cause-and-effect relationship. In fact, it looks as if population growth is lagging the emissions rather than leading them.

So what happened? Well, one scenario is that the growth in fossil fuel wealth in the West actually accelerated the population explosion in the developing world. Rising Western wealth exposed many parts of the developing world to some of the trappings of modern western life, such as minor modernisation and basic medicine. These factors led to increased reproduction and survival rates in the developing world. But this was without the full suite of cultural, societal and environmental factors that came into effect to constrain population growth rates in the West.

This recent world population growth has occurred mostly in developing nations: countries with very low fossil fuel consumption and emissions. So while most of the cumulative carbon emissions since the industrial revolution have occurred in the West, the sheer size of the Non-West's population means it now accounts for 70% of the world's GHG emissions. If the Non-West's CO_2e emissions (which are currently only 4.8 tCO2e per capita) increase by only 2 tCO2e per capita, it will negate the West eliminating all of its current CO_2e emissions.

Many in the emissions and climate change debate say the emissions problem is solely the result of overpopulation. This is unlikely, as population growth in the developing world may not be the cause of the rise in global emissions but rather an indirect result of it. But regardless of how we got here, it's hard to imagine any solution to emissions being achievable without an immediate turnaround in population growth. In this respect, whatever the cause may have been is irrelevant. Continued population growth is likely to counter any emission reduction initiatives we can put into effect.

This is part one of the Carbon Collision Course we are on.

CHAPTER 2

How fossil fuels and carbon emissions work

This chapter talks about the relationship between carbon dioxide emissions and fossil fuel consumption. It includes a simplified explanation of the chemistry that occurs during hydrocarbon fuel combustion. However, you don't need to understand hydrocarbon chemistry to understand emissions.

Here are two rules of thumb that are useful in quantifying CO_2 emissions from fossil fuel consumption:

1. Every kilogram of carbon burnt emits 3.67 kg of CO_2.
2. Hydrocarbon fuels are around 85% carbon by weight (on average), which is about the same as the average carbon content of crude oil and refined liquid fuels (petrol and diesel).

So on average, a kilogram of hydrocarbon fossil fuel emits 3.1 kg of carbon dioxide gas (0.85 x 3.67).

And a litre of petrol, which weighs about 0.75 kg and is 85% carbon by weight, emits about 2.34 kg of CO_2 (0.85 x 0.75 x 3.67).

The Carbon Cycle

All fossil fuels originate from organic (living) materials produced via the natural carbon cycle. They're derived from carbohydrates (materials containing carbon, hydrogen and oxygen), all of which originated through the process of photosynthesis in plants.

Here's a summary of the natural carbon cycle:

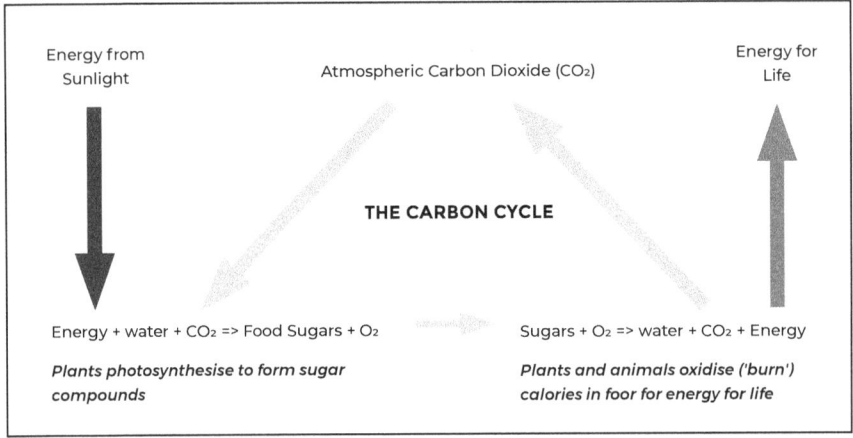

Of course, this is a simplification. Plants also use the captured energy to produce other organic plant materials (fats, proteins, cellulose, etc.) And animals also use the consumed energy to produce similar organic materials (fats, proteins, etc.). However, these compounds all eventually break down in accordance with the basic carbon cycle.

When a plant or animal dies in its natural environment, it's usually digested by other organisms. If it doesn't get eaten straight away, it will eventually rot or decay through digestion by micro-organisms. When this occurs, most of the organic carbon it contained is released back into the atmosphere as carbon dioxide. This means that in most cases an organism is still part of the natural carbon cycle even after it dies.

Fossil fuels originate from organic materials in plants or micro-organisms that weren't fully digested by other organisms. When the plants or micro-organisms died, they fell into naturally occurring low-oxygenated environments – lake beds, sea beds or bogs. The materials rotted anaerobically (without the presence of oxygen) at a slower rate than they would have during normal aerobic decay. In some cases, before the material fully decayed it was covered in layers of mud sediment that was eventually compressed into rock. Some material sunk deeper into the Earth's crust through further sedimentation and other geological effects. The combined effects of only partial decay and the heat and pressure of being buried in the Earth's crust transformed the material into what we know as the fossil fuels: crude oil, natural gas and coal.

While they originate from carbohydrate materials, fossil fuels predominantly consist of carbon and hydrogen in combination (hydrocarbons). A key effect of removing most of the oxygen from the original organic material is it concentrates the fuel's calorific energy content to a higher level than the original organic material.

The typical energy contents (known as Calorific Values or CVs) of fuels are:

Natural Gas	*52 MJ/kg (megajoules per kg)*
Crude Oils	*44 MJ/kg*
Coal	*36 MJ/kg (moisture- and ash-free basis)*
Wood	*15 MJ/kg (air-dried aged, 20% moisture)*
Sugar	*16 MJ/kg*

It's the high energy density of liquid petroleum fossil fuels that makes them so useful to human industrial activity. They have a higher energy density than the raw organic carbohydrate materials produced by natural photosynthesis.

The extraction and burning of fossilised fuels by humans is returning carbon previously stored in the Earth's crust back into the carbon cycle. While humans have been consuming fossil fuels in low quantities for centuries, in the past 175 years (since the modern industrial revolution) it has really taken off. Humans are currently consuming fossil fuels thousands of times faster than they were formed. The concern is that this, in combination with deforestation of the natural world, may be putting the natural carbon cycle out of balance and causing atmospheric CO_2 levels to rise. This in turn may be causing global warming.

I say *may* because we can't exactly measure what part fuel emissions are playing in the natural carbon cycle. As I mentioned earlier, fossil fuel emissions are estimated at 33 BtCO$_2$ per year. In comparison, the circulation rate of CO_2 in the natural carbon cycle is estimated to be about 800 BtCO$_2$[18]. So fossil fuel emissions factor in at only around 4% of the estimated natural CO_2 circulation rate.

But as I said in the disclaimer (on page ix), this book isn't about whether or not fossil fuels are raising atmospheric CO_2 levels, and therefore causing global warming. It's purely about the science of carbon emissions, and the political options to mitigate them.

Calories, kilojoules and kilowatts

To understand fossil fuel consumption, it's worth knowing a bit about how we consume energy. The last section compared the energy content of sugar (glucose) with gas, oils and coal. While food fuels (carbohydrates) and fossil fuels (hydrocarbons) are slightly different chemically, the way they're consumed to release energy is essentially the same. What's more, the energy content of food and fuels can be both expressed in calories or kilojoules.

The release of useful energy from food or fossil fuel is achieved by oxidising (reacting with oxygen) the carbohydrate or hydrocarbon material. Food is oxidised chemically in the body of living organisms to produce biochemical energy for life. The by-products of this process are the original compounds that went into formation of the organic material (by photosynthesis), water and carbon dioxide.

Food energy is used to power organisms' life support systems, such as – cell metabolism, cell renewal and growth etc. Most of the food energy consumed ends up back in the atmosphere as waste heat. This is either through emitted body heat, or warm air breathed out. Around 20% is used to produce useful work, generally in the form of physical activities required to support life such as collecting food and building shelter.

Fossil fuels are oxidised thermally by burning them in air. The products of this combustion are the same as living organisms – carbon dioxide, water and energy. However, in this case the energy produced is heat. This heat energy can be used directly or converted to either motive energy in an engine (such as an internal combustion engine) or electrical energy in a power station. Both types of machines are called 'heat engines' in thermodynamics (the science of energy transformation).

As with a living organism converting food energy into work, fuel energy can't be converted into motive or electrical energy with 100% efficiency. The efficiencies of various conversion processes range from 20% to 50%. The maximum theoretical efficiency a heat engine can achieve is about 55% for conditions on Earth as set by the laws of thermodynamics. Many of the machines for converting fuel into power are already pushing this limit.

The calorie content of food is usually expressed as food calories or kilojoules per 100g. This can also be converted and expressed as kilojoules per kilogram. When the energy content of food is expressed this way it's exactly comparable to the energy content of a fossil fuel. One food calorie is equal to 4.2 kJ, so the calorie content of pure sugar (380 Cal/100g) is equivalent to 16 kJ/g or 16 MJ/kg (megajoules per kg).

The term 'power' refers to the rate work energy is produced or consumed, and is measured in watts (W) or kilowatts (kW). One watt of power is work energy used at one joule per second. The kW unit can also be used to express the rate of calorific energy consumed.

An average human in the West with an office-based/low-labour-intensive job might consume about 2,500 calories of food energy per day. This energy consumption rate can be converted to kW as follows:

2,500 x 4.2 = 10,500 kilojoules per day (kJ/day)

10,500 / 24 / 60 / 60 = 0.122 kilowatts (kW, or kJ/sec)

An average human consuming 2,500 calories a day is consuming energy at a rate of about 0.122 kW.

In Australia, fossil fuels are consumed at an average rate of 7.2 kW per capita. In other words, we consume fossil fuel energy at about 60 times the rate at which we consume food energy.

The kilowatt-hour (kWh)

While not strictly a unit of measure in the metric system, the kilowatt-hour is the primary unit of measure for electrical energy production and consumption. However, it can be used to measure any form of energy. As mentioned earlier, the watt or kilowatt is the unit of measure of power or rate of energy consumption or production. With this in mind, the kilowatt-hour (kWh) term may seem confusing. But the kWh isn't a unit of power. It's a unit of energy. More specifically, it's the amount of energy equivalent to one kW of power being produced or consumed continuously for one hour.

A kilowatt is equivalent to 1 kJ/sec. And as there are 3,600 seconds in an hour, 1 kWh is equivalent to 3,600 kJ of energy.

A continuous power consumption of 1 kW will consume:

1 kWh of energy over an hour
24 kWh of energy over a day
8,760 kWh of energy over a year (24x365)

How we purchase and consume energy and power

A consumer's directly attributable CO_2 emission is primarily made up of three types of energy and power consumption:

1. Electrical power generated from fossil fuels, coals, petroleum oil fuels or natural gas

2. Petroleum fuels for transportation, including airline travel

3. Petroleum fuels or natural gas for domestic heating and cooking.

Non-directly attributable consumer CO_2 emissions are those associated with supplying and purchasing other goods and services not directly associated with energy consumption. It is safe to assume that almost every purchase made consumes some form of fuel or energy to either produce or transport it, and will therefore have an associated carbon footprint.

A note on carbon footprint calculators

A number of personal carbon footprint calculators are available on the internet, and they all work in a similar fashion: the consumer enters their consumption data, and the calculator applies typical emission factors to calculate an estimated CO_2e footprint.

These calculators typically consider:

- Home electricity usage (kWh/day)
- Vehicle type and usage (km/year)
- Natural gas usage (generally heating)
- Air travel use
- Public transport use.

Some calculators also estimate emission footprints for particular eating habits. For example, beef and dairy products result in higher emissions of methane (a greenhouse gas with a global warming potential 25 times that of CO_2) in the agricultural sector. Methane emissions result from so-called enteric fermentation in the digestive systems of ruminant animals – cows, sheep, goats, etc. (Contrary to popular belief, most methane is emitted from the animal belching rather than from flatulence).

The problem with individual footprint calculators is they don't consider the economy as whole.

For example:

- Emissions resulting from providing government services. This may include subsidies for underutilised public transport in off-peak periods (empty buses and trains).
- Emissions resulting from producing goods for export, including natural resources and fossil fuels.

This book focuses on per capita emissions, but considers emissions generated by the whole economy –not individual consumers.

Emission factors and emission intensities

As shown in the previous section, fossil fuels are predominantly made up of compounds of carbon and hydrogen in combination (generally referred to as hydrocarbons). Oils and coals can have thousands of types of hydrocarbon molecules. Fortunately, you don't need to know the fuel's exact chemical composition to estimate how much CO_2 it emits when it combusts.

All hydrocarbon molecules combust or burn in air according to this general chemical formula:

$$C_aH_b \quad + \quad a+b/4\ O_2 \quad \rightarrow \quad a\ CO_2 \quad + \quad b/2\ H_2O$$

Hydrocarbon	Oxygen		Carbon Dioxide	Water

As you can see, every atom of carbon burned produces one molecule of carbon dioxide, regardless of the starting compound's exact chemical composition. Carbon's atomic weight is 12 and oxygen's is 16, so the molecular weight of carbon dioxide is 44 (12 + 16 x 2).

Which means every 12 kg of carbon burned will always produce 44 kg of carbon dioxide.

To estimate the CO_2 emission associated with a particular fuel, all we need is its approximate weight percent (%-wt) of carbon. And these are available for all fuel types.

The ratio of CO_2 emitted to carbon weight % in the raw fossil fuel will always be 44/12 (= 3.666 recurring, rounded in here to 3.67).

Every tonne of carbon we burn generates 3.67 (44/12) tonnes of carbon dioxide.

The three main fossil fuel classes are natural gas, oils and coals. Although oils and coals from different source reservoirs can have different chemical compositions, the carbon content of their hydrocarbon component is similar enough that we can summarise them as follows:

	Wt % Carbon in Hydrocarbon	$kgCO_2/kg$
Natural Gas	75%	2.75
Crude Oils	85%	3.12
Coals	95%	3.48

As you can see, the range of weight percentage of carbon in hydrocarbon is relatively narrow, from 75% in natural gas to 95% in coals. The resulting emissions range from 2.75 tCO_2 to 3.48 tCO_2 *per tonne of hydrocarbon.*

However, there's a complication. Processed natural gas and liquid petroleum fuels are nearly pure hydrocarbon in their refined forms. But coals are rarely pure hydrocarbons. They contain quantities of other elements such as some remaining combined oxygen, nitrogen, and in some cases sulphur. They also contain quantities of non-combustible materials (known as ash) and varying quantities of water. Coals vary in hydrocarbon concentration from 25% for wet brown coals to 90% for anthracite grades. Typical compositions used here are 28% for Australian brown coal (lignite) and 64% for Australian black coal (bituminous coal).

To estimate the quantity of CO_2 generated from burning each fuel type, the purity of its hydrocarbon component needs to be taken into consideration. This is shown in the following table.

Hydrocarbon fuel	wt% C in HC	wt% HC	EF $kgCO_2$/ kg-Fuel
Natural gas	75%	98%	2.70
Fuel oil	87%	98%	3.13
Brown coal	93%	28%	0.96
Black coal	94%	64%	2.21

The amount of CO_2 emitted when using a particular fuel is termed its emission factor (EF). And so from the previous discussion the formula for emission factor is:

$$EF \quad = \quad \text{Weight \% Carbon in the fuel} \quad x \quad 44/12$$

Where: EF is in $kgCO_2$/kg fuel

As you can see, crude-oil-derived fuel oil has the highest CO_2 emission factor, followed by natural gas and black coal, with brown coal having the lowest emission factor.

This contradicts the popular view that natural gas is 'cleaner' (in terms of CO_2 emission) than coal. The reason for this apparent anomaly is the emission factor doesn't take into account the amount of useful energy the fuel provides. The emission of CO_2 per unit of work energy delivered is termed the emission intensity (EI) of the fuel.

Emission intensities for electricity generation

As shown in the previous section, the emission factor (EF) is the amount of CO_2 released by burning a specific amount of a fuel. The emission intensity (EI) is the amount of CO_2 released by burning enough fuel to generate a specific amount of electricity or motive work. (Electricity and motive work energy can be considered equivalent for this level of analysis).

The two major processes that convert fossil fuel energy into useful work are:

1. Rotational power generation in internal combustion engines, i.e. motor vehicles etc.

2. Electricity generation in power stations.

The emission intensity is derived from the emission factor by considering:

*   The calorific value of the fuel

*   The efficiency of converting the fuel to useful work.

As useful work is generally quoted in the same units of measure as electricity (kWh), emission intensity is also generally quoted in units of $kgCO_2/kWh$.

The equation to convert emission factor to emission intensity is:

$$EI = \frac{EF \times 3.6}{(CV \times Efficiency)}$$

Where: Efficiency - refers to the efficiency of the conversion of heat energy to work energy in a heat engine

CV is the calorific value in MJ/kg

3.6 is the factor to convert from $kgCO_2/MJ$ to $kgCO_2/kWh$

EI is in $kgCO_2/kWh$

Here are the results for the main fossil fuel type grouping (using electricity generation efficiency assumptions).

Hydrocarbon fuel	EF $kgCO_2/$ kg fuel	Cal Value MJ/kg	Efficiency (Typical)	EI $kgCO_2/$ kWh
Natural gas [1]	2.70	52.5	48%	0.39
Fuel oil [2]	3.13	44.8	40%	0.63
Brown coal [3]	0.96	10.2	31%	1.10
Black coal [3]	2.21	24.4	38%	0.86
Black coal HELE [4]	2.21	24.4	43%	0.76

Notes:
1. *Natural gas power generation in combined cycle gas turbine (CCGT) power plant.*
2. *Fuel oil in conventional steam turbine (ST) power plant.*
3. *Brown and black coal in conventional steam turbine (ST) power plant.*
4. *Black coal in so-called high-efficiency, low-emission (HELE) steam turbine (ST) power plant.*

As you can see, while black and brown coals have lowest emission factors (EF) they have the highest emission intensities (0.9 and 1.1 $kgCO_2/kWh$ respectively). This is due to the combination of lower calorific values and lower efficiencies in coal-fired power generation facilities.

Natural gas power generation (in high-efficiency CCGT facilities) has an emission intensity of around 0.4 $kgCO_2/kWh$. So CO_2 emissions from natural gas power generation are about 40% that of coal on a $kgCO_2/kWh$ basis (averaging black and brown).

High-efficiency, low emission (HELE) coal

At the time of writing, the Australian coal industry was promoting increased use of coal for electricity generation in Australia[20]. The website promotes the use of so-called high-efficiency, low-emission (HELE) coal fired power generation, also referred to as 'Clean Coal'.

HELE technology uses various techniques to improve black coal fired generation efficiency, such as:

•   Pre-pulverisation of the coal to improve combustion efficiency

•   Higher temperature (also known as "super-critical") steam to improve turbine generator efficiency.

To quote the website:

> *High-efficiency, low emission (HELE) technology enables coal-fired power stations to operate at higher temperatures and pressures, delivering electricity more efficiently and reducing emissions by up to 40%[†]. HELE is already making a difference at 4 power stations in Queensland.*
>
> [†] *ACA Low Emissions Technologies assessment based on publicly available information on world power plant efficiency levels, July 2015.*

The efficiency gain for high-quality black coal HELE generation is an improvement in absolute terms to between 40% and 45% – not an efficiency improvement of 40% to 45%.

This 40% emission reduction claim compares HELE generation to the average worldwide brown and black coal-fired generation. It does *not* compare a similarly designed black coal conventional power generation facility with and without HELE technology.

The previous table shows HELE technology black coal power generation at an assumed efficiency of 43%, versus a typical non HELE efficiency of 38%.

HELE technology only lowers emission by about 12% when compared to a similarly designed black coal conventional power generation facility without HELE technology. The improvement in emission performance is in the same ratio as the improvement in efficiency, which is 43%/38%. It doesn't lower emissions to the level of natural gas.

HELE in itself won't significantly lower emissions from coal-fired power generation. And it certainly won't achieve a 40% reduction when compared to a modern facility without HELE technology. It might be better to term it as 'higher-efficiency, lower-emissions' technology. It might also be better to describe it as "cleaner" coal rather than "clean" coal.

Liquid petroleum fuels emission factors and intensities

In the same way CO_2 emission factors were estimated for the raw material fuels used for electricity generation, similar estimates can be made for liquid transportation fuels: petrol, diesel, ethanol-blended petrol and LPG. For liquid transportation fuels, it's useful to express the emission factor as kg CO_2 per litre of fuel, or $kgCO_2/L$. This can be done by multiplying the EF in $kgCO_2/kg$ fuel by the average density of the fuel in kg/L.

Hydrocarbon fuel	wt% Carbon	EF $kgCO_2/$ kg fuel	Density kg/L	EF $kgCO_2/$ L fuel
Liquefied petroleum gas (LPG)	82%	3.02	0.542	1.63
Petrol – average	85%	3.12	0.750	2.34
Petrol E-10 [1] [2]	81.7%	2.79	0.749	2.09
Aviation fuel	85%	3.12	0.800	2.49
Diesel	86%	3.15	0.840	2.65

Notes:
1. *E-10 is petrol blended with 10% ethanol derived from plant material.*
2. *CO_2 from the 10% ethanol component not included in the CO_2 emission for E-10 petrol.*

The table shows that diesel has the highest CO_2 emission factor at 2.65 kgCO_2/L. There isn't much difference in carbon content between diesel and petrol. (The difference is driven by the higher density of diesel.) Liquefied petroleum gas (LPG) has the lowest emission factor at 1.63 kgCO_2/L. This is driven by both lower carbon content and lower density.

However, just because diesel has the highest emission factor doesn't necessarily mean a diesel-engine vehicle will emit more CO_2 than a similarly designed and weighted petrol engine vehicle. In the way calorific-to-electrical conversion efficiencies were used to convert emission factors to emission intensities for electricity generation fuels, the same has to be done for transportation fuels to make a fair comparison

This has been done in the following table, which also shows the estimated fuel economy of each fuel in kWh/L.

Hydrocarbon fuel	EF kgCO₂/ L fuel	Cal Value MJ/kg	Efficiency (Typical)	EI kgCO₂/ kWh	Fuel Economy kWh/L
LPG	1.63	50.0	27%	0.805	2.03
Petrol – average	2.34	47.3	27%	0.879	2.66
Petrol E-10	2.09	45.4	27%	0.818	2.55
Aviation fuel	2.49				
Diesel	2.65	44.8	30%	0.845	3.14

Diesel

This table shows that while diesel has a higher emission factor than petrol, it actually has a lower emission intensity That's because diesel has the highest fuel economy of any of the fuels due to higher fuel density and engine efficiency. For a similarly designed and weighted car under similar operating conditions, diesel delivers about 18% better fuel economy than petrol. The combination of emission factor and improved fuel economy means diesel emits about 95% of the CO_2 emitted by petrol.

E-10 Petrol - 10% ethanol blended petrol

Using E-10 results in a CO_2 emission of about 94% of the emission from regular petrol. The reduction assumes that all the direct CO_2 emitted by ethanol is renewable. (This doesn't take into account the fossil fuel emissions associated with ethanol manufacture). Without this assumption, the CO_2 emission from E-10 fuel is almost identical to regular petrol.

The calorific value of pure ethanol is only 66% that of petrol. When blended at 10%, it results in a blend with a calorific value of about 97% that of petrol. This means the consumer needs to purchase about 3% more E-10 to get the same amount of useful energy (e.g. distance travelled). So although E-10 petrol does reduce emissions (if the assumption about ethanol being 100% renewable is correct), E-10 petrol needs to be about 3% cheaper to make the consumer cost neutral to the decision to use it.

If all petrol sold in Australia was mandated to be E-10 (and again assuming the ethanol component is 100% renewable), it would result in a CO_2 reduction of about 2.7 million tCO_2/year (0.1 tCO_2 per capita). However, this is only about 0.5% of Australia's total emission, and so using E-10 petrol appears to have only a limited capability of reducing Australia's carbon footprint.

Liquefied petroleum gas (LPG)

LPG has the lowest fuel economy of the liquid fuels, at only 76% that of petrol. But as it generally retails at a big enough discount to balance this, using LPG doesn't incur a cost penalty compared to using petrol. Taking into account the low emission factor, LPG has the lowest CO_2 emission intensity at around 92% that of regular petrol.

NOTE

Emission Intensities (EI) estimates in this chapter compare the generic fuel and emission performance of a particular fuel and engine configuration.

Unlike power generation, the power (or work) output of a motor vehicle engine isn't metered and tracked. The best way to estimate actual CO_2 emissions for a specific vehicle is to measure the actual litres consumed and multiply that figure by the emission factor of the fuel in $kgCO_2/L$. For example, using ten litres of petrol results in an emission of 23.4 $kgCO_2$.

Emissions from Air Travel

Consumption of aviation fuel contributes about 24 $MtCO_2$/year to Australia's reported GHG emissions. This is around 1 tCO_2 per capita, or about 4% of total emissions.

Aviation fuel has an average carbon content of about 85% and an average density of 0.8 kg/L, resulting in an emission factor of 2.49 $kgCO_2$ per litre of aviation fuel consumed.

Compared to car travel, economy-class air travel is relatively efficient in terms of litres of fuel consumed per 100 km travelled per passenger (L/100 P-km). Typical fuel consumption for various flight types are:

Short haul	*3.2 L/100P-km*
Medium haul	*4.3 L/100P-km*
Long haul	*4.6 L/100P-km*

Unlike most transportation fuel use, average fuel consumption and emissions during air travel do not decrease as distance travelled increases. This is due to the extra fuel that needs to be carried at take-off and during the early part of the flight. Unlike a road vehicle, an aircraft needs to expend part of the lift power it generates to hold up the weight of the fuel it's carrying. (This is termed *sacrificial load*.) Fuel consumption for an aircraft is a function of its total weight, so the higher weight of fuel that needs to be carried in the early part of the flight impairs the average fuel economy for long-distance flights.

Although air travel is relatively efficient in terms of L/100 P-km, that doesn't mean it's a low-emission activity. Air travel allows us to travel much further than we can by a motor vehicle. If minimising emissions was the criteria for taking a particular holiday, it would be better for a single passenger to drive for ten hours in a typical car than to take a typical ten-hour flight.

	Car	Flight
L/100 P-km – typical	10.0	4.6
Average speed km/h	90	900
L/h	9	41
Time h	10	10
Distance km	900	9,000
L used	90	414
EF	2.35	2.49
tCO$_2$ per trip	0.21	1.03

And travelling with multiple occupants would result in even lower emissions per person.

Here are the CO$_2$ emissions for various flight typical destinations:

Air travel	kg CO$_2$/ L-fuel	L/100 P-km	Dist. Km – one way	L Fuel per Pass.	tCO$_2$ per Pass.	Example
Short haul	2.49	3.2	1,000	32	0.08	Brisbane to Sydney
Medium haul	2.49	4.3	7,500	323	0.80	Brisbane to Bangkok
Long haul	2.49	4.6	17,000	782	1.95	Brisbane to London

A single-leg flight from Brisbane to Sydney would emit about 80 kgCO$_2$ per passenger, while a single-leg flight from Australia to Europe would emit about 1.95 tCO$_2$ per passenger.

Air travel vapour trail Greenhouse effect

However, these estimates are only for direct CO_2 fuel combustion emission. It is estimated that high-altitude combustion and vapour trails from aircraft double the effective CO_2e (equivalent) emission over the base fuel CO_2 emission[2]. This isn't generally taken into account in the CO_2 emissions used to estimate carbon offset requirements in airline offset programs. (See Chapter 8.)

Business class travel

Airline emissions data sources generally don't differentiate between business and economy class travel.

The Victorian Government EPA carbon emissions calculator[23] estimates business class emission to be:

> *1.8 times economy base CO_2 emission for domestic business class*
>
> *3.4 times economy base CO_2 emission for international business class (sleeper).*

This means a single-leg business class trip from Australia to Europe would emit about 6.63 tCO_2 per business class passenger in just 24 hours. That's more than a whole year's per capita emissions for the 6.4 billion people living in the Non-West (4.8 tCO_2e).

Carbon tax – theoretical impact on energy prices

Although the carbon tax in Australia was repealed in 2014, it's worth knowing the potential impact a theoretical carbon tax would have on fossil fuel prices for electricity and transportation fuels.

We can evaluate the effect on energy and fuel cost by multiplying the carbon price (in $/kgCO_2$) by the applicable emissions intensities (for electricity) or emission factors (for liquid fuel).

A carbon tax effectively provides an incentive to reduce CO_2 by discouraging fossil fuel consumption. At what level it should be set to be effective is a key question. It's generally accepted that a carbon price of around 100 \$/tCO2 is needed to make carbon capture and storage (CCS) economically viable. So let's use an example price of \$100/tCO2 (which is 0.1 \$/kgCO2) to illustrate the effect of a theoretical carbon tax.

This table shows the effect of \$100/t carbon price/tax on electricity generation fuel costs:

Hydrocarbon fuel	Effect of $100/tCO_2 Carbon Tax	Units
Natural gas	3.9	c/kWh
Fuel oil	6.3	c/kWh
Brown coal	11.0	c/kWh
Black coal	8.6	c/kWh

And this table shows the effect of \$100/t carbon price/tax on liquid petroleum fuel costs:

Hydrocarbon fuel	Effect of $100/tCO_2 Carbon Tax	Units
LPG	16.3	c/L
Petrol – average	23.4	c/L
Petrol E-10	20.9	c/L
Aviation fuel	24.9	c/L
Diesel	26.5	c/L

Effect of a carbon tax on air travel

Consumers generally aren't informed of the fuel consumption associated with their air travel activities. However, we can use the emissions estimated in the previous section to estimate the effect of a $100/tCO_2 carbon price on air travel on example routes.

Hydrocarbon fuel	Dist. Km – One way	Business Class Factor	L/100 P-km	tCO_2	Effect of $100/ tCO_2 Carbon Tax $
Short haul – economy	1,000		3.2	0.08	8
Medium haul – economy	7,280		4.3	0.78	78
Long haul – economy	17,000		4.6	1.95	195
Short haul – business	1,000	1.8	5.8	0.14	14
Medium haul – business	7,280	3.4	14.6	2.65	265
Long haul – business	17,000	3.4	15.6	6.63	663

Note: *This excludes the estimated doubling greenhouse gas effect of vapour trails from air travel. If this was taken into account, these costs would also double.*

CHAPTER 3

Australia's emissions profile

The Australian Federal Government calculates greenhouse gas (GHG) emissions for Australia each quarter using data gathered under reporting obligations for all business and organisations. The results are then published in a set of reports called Australian National Greenhouse Gas Inventory (NGGI) Reports, published by the Department of the Environment and Energy[3].

Here's a summary of Australia's NGGI reported GHG emissions for 2015.

Emission Sector	MtCO₂e 2015 NGGI	tCO₂e per Capita	% of Total
Fossil fuel usage	375.0	15.8	71%
Fugitive emissions (1)	44.5	1.9	8%
Industrial processes	32.3	1.4	6%
Agriculture	70.0	3.0	13%
Waste processing	11.4	0.5	2%
Total excluding LULUCF	**533.2**	**22.5**	101%
LULUCF	-7.7	-0.3	-1%
Total including LULUCF	**525.5**	**22.2**	100%
Population - millions	23.7		

Note: *Fugitive emissions are escaping methane (natural gas) associated with fossil fuel extraction and use. Source: [3].*

The table makes a number of references to LULUCF—Land Use, Land Use Change and Forestry. These are net CO_2e emissions resulting from Land Use Change and Forestry activities. The term 'net' is used because they can also include carbon calculated to be captured by land change activities such as reforestation. The significance of land use change sector emissions in Australia's emission profile is explained later in the chapter.

The officially reported GHG emissions for Australia for 2015 were 525.5 million tonnes of carbon dioxide equivalent (tCO_2e). The population estimate at the time was 23.7 million, resulting in a calculated per capita emission of 22.2 tCO_2e.

The most significant contributor to Australia's GHG footprint is fossil-fuel-based energy consumption. 79% of the CO_2e emissions are associated with producing and consuming fossil fuels for electricity, transportation, or other uses such as heating. 90% of this is the direct emission of actual carbon dioxide (CO_2) in the combustion of the fuel. That's about 16 tCO_2e per capita year, or 71% of the GHG footprint.

Other government publications can be used to estimate the breakdown of the fossil fuel consumption contribution to the CO_2 emissions reported in the reports. This analysis is shown in the following sections.

Liquid petroleum fuel emissions

The Department of the Environment and Energy publish estimates of total liquid petroleum fuel sales in Australia. For 2015, approximately 55,000 million litres of liquid fuels were sold, as detailed in the table below. The CO_2 emission factors estimated in the previous chapter have been applied to estimate the emission associated with the consumption of each fuel.

Petroleum fuel	ML 2015/6	L per capita	EF kgCO2/ L	tCO2 per capita
LPG	3,200	135	1.63	0.22
Petrol	16,275	687	2.34	1.61
Petrol E-10 blended (Ethanol)	1,971	83	2.10	0.17
Diesel fuels	23,866	1,007	2.63	2.65
Aviation fuel (domestic)	3,942	166	2.49	0.41
Aviation fuel (international)	4,569	193	2.49	0.48
Fuel oil	862	36	3.13	0.11
TOTAL	54,685	2,307	2.45	5.66
Population	23.7			

Source: [8]

In summary:

- The total consumption of liquid petroleum fuels in Australia is about 2,300 litres per capita year.

- The per capita CO_2 emission from petroleum fuel usage is estimated at 5.7 tCO_2 per year.

Note that while this has been calculated as a per capita figure, the fuel consumptions reported are across all aspects of the economy (retail, commercial, industrial, government, etc.).

Electricity generation emissions

The Australian Energy Update[9] provides estimates of Australia's yearly on-grid electricity production. For 2014/5 this was around 252,000 million kWh, which is equivalent to continuous electrical power consumption of about 1.2 kW per capita. The report also provides a breakdown of electricity generation by source. Again, the CO_2 emission intensities for each generation fuel estimated in the previous chapter have been applied in the following table:

Generation fuel	M-kWh 2014/5	% of Total	EI kgCO$_2$/ kWh	tCO$_2$ per capita
Black coal	107,639	43%	0.86	3.89
Brown coal	50,970	20%	1.10	2.36
Gas (CCGT)	52,463	21%	0.39	0.85
Fuel oils	6,799	3%	0.63	0.18
Hydroelectricity	13,445	5%		
Wind	11,467	5%		
Solar PV	3,608	1%		
Bio-generation	5,968	2%		
Geothermal	1	0%		
TOTAL	252,360		0.691	**7.29**
Population	23.7			
kW per Capita	**1.22**			

Source: [9]

In summary:

1. The total consumption of electricity in Australia is about 1.22 kW per capita.

2. The per capita CO$_2$ emission from electricity usage is estimated at 7.3 tCO$_2$ per year.

Again, while this has been calculated as a per capita figure the electricity usage reported is from across all aspects of the economy.

Internal natural gas and refining fuel consumption emissions

The other main source of fossil fuel CO$_2$ emission is from internal consumption of natural gas for uses other than electricity (industrial, domestic heating, etc.), as well as internal fuel consumption in natural gas and oil production operations, and oil refining operations.

Processing natural gas for domestic consumption (either electricity or industrial/domestic heating) consumes a portion of the gas it produces as internal fuel. Similarly, crude oil refining consumes about 6% of the fuel it produces as internal fuel.

Here are the estimates for these consumptions and emissions:

Natural gas - all other uses [1]	Mt Produced 2014/5	% used as Fuel	Mt Fuel 2014/5
Gas - other uses	16	100%	15.8
Gas production	50	6%	3.0
Gas production - electricity	10	6%	0.6
Export LNG	24	14%	3.4
Refining fuel usage [2]	22	6%	1.3
TOTAL	**121**		**24.1**
Population	23.7		

Notes:
1. *Includes all inland use of natural gas. It doesn't include gas directly used for electricity generation, but does include fuel consumed in processing gas for electricity generation.*
2. *Estimate for inland oil refining internal fuel use (~1 Mt/year natural gas equivalent).*

Natural gas - other uses total	Mt 2014/5	t per capita	EF kg-CO_2a/ kg	tCO_2 per capita
Natural gas equivalent	24.1	1.01	2.7	2.7
Population	23.7			

Sources: [8], [9]

In summary:

1. The total consumption of natural gas in Australia is about 1 t per capita year.

2. The per capita CO_2 emission from natural gas usage is estimated at 2.7 tCO_2 per year.

Summary

The combined per capita CO_2 consumption and emissions for the various fossil fuel used in Australia are summarised as follows:

Fossil fuel usage	Units	Use per capita	tCO₂ per capita	kW fuel per capita
Liquid Petroleum Fuels	Litres	2,307	5.66	2.6
Electricity	kW	1.22	7.29	2.9
Gas and Refining	Tonnes	1.01	2.71	1.7
TOTAL			**15.7**	**7.2**

This is in close agreement with 15.8 tCO₂ per capita calculated from the 2015 NGGI reported emission for hydrocarbon fuel use.

Australia's emissions compared with other counties and country groups are discussed in Chapter 4.

Australia's emissions performance 1990–2016 – The Kyoto Protocol and Paris Targets

The Australian Government makes the following claim in relation to emissions performance under the first round of GHG emission targets under the Kyoto Protocol[21]:

> *"Our targets build on our success to date*
>
> *Australia has a proud history of meeting and beating our international commitments on climate change.*
>
> *Australia out performed its first target under the Kyoto Protocol. Our Direct Action Plan on climate change has us on track to meet our commitment to reduce emissions by five per cent below 2000 levels by 2020, which is equivalent to 13 per cent below 2005 levels."*

The following chart shows Australia's GHG emissions during the period 1990 to 2016, produced from data contained within various NGGI Reports.

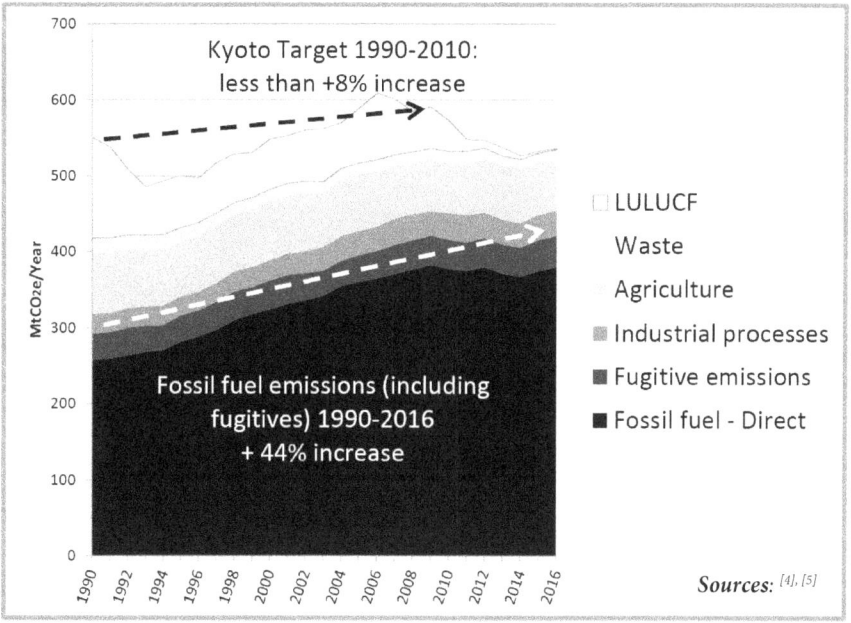

The chart and supporting data demonstrate the following outcomes:

1. Overall GHG emissions in Australia have reduced slightly from 551 MtCO₂e in 1990 to 525 MtCO₂e in 2015, a reduction of 5%.

2. The only emission sector to show a significant reduction over the period is the Land use change (LULUCF) sector, which has reduced its emissions from 132 MtCO₂e to -7 MtCO₂e.

3. Fossil fuel emissions, including fugitive methane emissions (escaping or leaking natural gas) has increased from 292 MtCO2e to 421 MtCO2e, which is a 44% increase.

4. Emission peaked at 610 MtCO₂e in 2006. Since then emissions have fallen back to 536 MtCO₂e. But again, the only significant reduction has been again in the land use change sector.

Kyoto Protocol round one target

Australia's target for the first round of Kyoto Protocol targets was for average emissions in 2008–2012 to not be more than 8% greater than 1990 levels. While most Western or developed countries had targets reducing GHG emissions, the Australian Government negotiated a target limiting emissions growth to no more than 8% in anticipation of a population growth rate higher than the Western average.

Australia easily met its Kyoto Protocol round one target. Average emissions in 2008–2012 were 569 MtCO$_2$e (a 3% increase), so Australia beat the target by around five percentage points. However, this may not result in an overall emission decrease as the government was also allowed to carry over any improvement on the first target as a relaxation to the second-round emission target.

The second round Kyoto Protocol target was for emissions in the period 2013-2020 to be 5% lower than emissions in 2000. Emissions in 2000 were 548 MtCO$_2$e. So 'carrying over' 5% means Australia already has a head start on its second-round target.

It was mentioned that the only significant emission sector to show reduction between 1990 and 2008–2012 was the land use change sector. This is explained in the next section.

LULUCF and 'The Australia Clause'

LULUCF, or land use change emissions, are estimates of net CO_2e emissions resulting in changes in the use of vegetated land use such as forests, grasslands and croplands. Like most carbon emissions, land use change emissions are calculated by accounting methods rather than being measured directly. Land use change emissions are tracked in the National Carbon Accounting System (NCAS), which relies on satellite data to track trends in land clearing.

The five categories of land use change used in Australia's accounts are:

1. Forest land remaining forest land

2. Land converted to forests

3. Forest converted to other uses

4. Cropland remaining cropland

5. Grassland remaining grassland (including wetlands and settlements)

'Forest converted to other uses' is the term generally used to cover deforestation (land clearing), and has been the most significant activity in Australia's calculated land use change emissions over the Kyoto Protocol round one target years:

- Net land use change emissions were 132 $MtCO_2e$ in 1990.

- "Forest converted to other uses" resulted in emissions of 248 $MtCO_2e$, primarily associated with 600,000 hectares of land clearing (6,000 km^2).

- Net land use change emissions in 2015 were -7 $MtCO_2e$.
  Source: [10]

When forest land is cleared, the wood is rarely used for wood products or fuel. Instead it is either burned off or set aside to rot. Either way, the carbon it contained is released back into the atmosphere as CO_2.

As well as negotiating an increase in emissions over the first Kyoto Protocol target period, Australia negotiated a clause into its GHG emission target to include land use change emissions in the baseline year of 1990. Although 1990 was the Kyoto Protocol target baseline year, the actual target negotiations weren't concluded until 1997. It was already known that land clearing in Australia peaked in 1990 with around 600,000 hectares of land being cleared, releasing a calculated net quantity of 130 $MtCO_2$ into the atmosphere. Thus 130 $MtCO_2e$ of land use change sector emissions were included in the baseline that future emission performance would be compared with.

This became known as 'The Australia Clause'. It meant that Australia's emissions from fossil fuel consumption could actually rise and still have its overall emissions increase come in at under 8%. As stated at the beginning of this section fossil fuel emissions, including fugitive methane emissions associated with fossil fuel extraction and use, have increased from 292 $MtCO_2e$ in 1990 to 421 $MtCO_2e$ in 2016, an increase of 44%.

Per capita emissions

The Kyoto Protocol round one target for Australia was to allow for population growth. So it's only fair to look at the trend on a per capita emissions basis, as shown in the following chart.

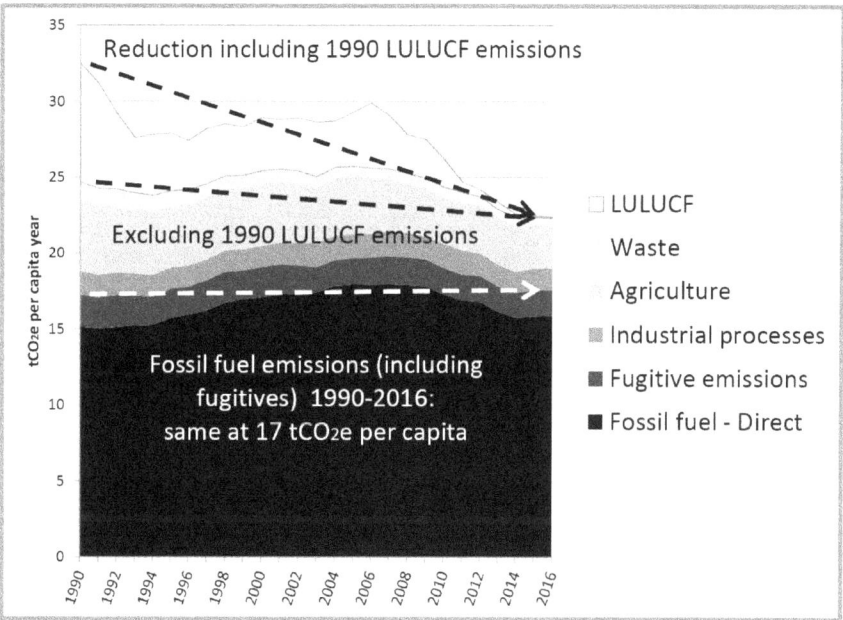

This chart does show a reduction in per capita emissions over the Kyoto target period and through to 2016. Per capita emissions have reduced from 32.5 tCO2e in 1990 to 22.2 tCO2e in 2016.

However, we should also note that:

- Again, most of the reduction in per capita emission is due to the land use change emission inclusion in the 1990 baseline.

- Excluding land use change, per capita emissions were reduced from 24.5 tCO2e to 22.2 tCO2e, a reduction of about 9%.

- Most of this reduction has been in the agriculture and waste processing emission sectors.

- Fossil fuel emissions, including fugitive methane emissions, are the same in 2016 as they were in 1990 – around 17.5 tCO₂e per capita.

- Fossil fuel emissions (including fugitive emissions) peaked at around 19 tCO₂e per capita in 2006.

So while Australia has reduced some emissions in sectors outside of land use change, emissions from the biggest emission source (fossil fuels) have remained almost constant over the Kyoto Protocol round one target period.

The Kaya Identity – linking emissions to Gross Domestic Product

The Kaya Identity[22] is an equation proposed to quantify the human emissions of CO_2 as a function of three key ratios used in combination with the population (Pop):

- Energy consumed per unit of Gross Domestic Product (En/GDP)

- Gross Domestic Product per capita (GDP/Pop)

- CO₂e emissions per unit of energy consumed (CO₂/En)

According to the Kaya Identity, total emissions can be expressed as:

Total Emissions = Pop x GDP/Pop x En/GDP x CO₂/En

A number of governments, including China and Australia, use a form of Kaya Identity to demonstrate progress on emissions reduction. They focus on the ratio CO_2/GDP, which is termed emission intensity of the economy.

Total Emissions = Pop x GDP/Pop x CO₂/GDP,
(Where CO₂/GDP = CO₂e emissions per unit of GDP)

Note: The economic emission intensity should not be confused with the fuel emission intensity discussed in Chapter 2.

For Australia, the economic emission intensity has reduced significantly during the Kyoto Protocol years. It has come down from 0.72 in 1990 to 0.32 in 2016, a reduction of 56%.

The following chart headlines the executive summary of the *Quarterly Update of Australia's National Greenhouse Gas Inventory: June 2017* report.

Figure P1: Emissions per capita and per dollar of real GDP (incl. LULUCF), financial year, 1990 to 2017

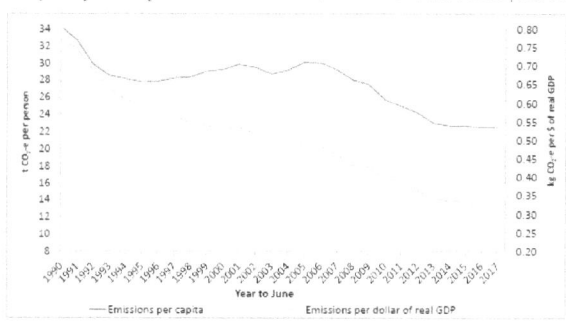

Source: Department of the Environment and Energy.

The commentary to the chart states:

> *"The emissions per capita and the emissions intensity of the economy were at their lowest levels in 28 years in 2016–17. Emissions per capita in 2016–17 have fallen 34.4 per cent since 1990, while the emissions intensity of the economy has fallen 57.8 per cent since 1990 (Figure P1)". (Source: [6])*

The first issue with this being evidence of good environmental performance is that the per capita emissions reduction claim includes land change use emission in the 1990 baseline ('the Australia Clause'). As shown previously, GHG emissions excluding land use change have reduced, but only from 24.5 to 22.2 tCO2e per capita. And there have been no reductions in the fossil fuel component of the per capita emission.

So what about the emission intensity reduction claim? Unfortunately, this doesn't actually mean anything tangible in terms of emissions reduction either. The main reason this indicator shows improvement is due to the increase in GDP.

The basic problem with the Kaya Identity is that all forms of the equation can be reduced to:

$$Total\ Emissions = Total\ Emissions$$

Bringing ratios involving GDP into the equation just muddies the waters. GDP generally increases due to factors not associated with energy consumption. It may even increase because of differing assumptions about inflation. For Australia, a key factor in GDP growth over the 1990–2016 period has been the near exponential rise of fossil fuel and mineral exports. So counting a reduction in emission intensity in $kgCO_2e/\$-GDP$ in Australia as a positive performance indicator in the global warming debate is a somewhat obtuse claim.

My view is that any references to GDP or cost or financial parameters in Kaya-type equations should be removed.

The only key performance indicator that matters in quantifying GHG emissions progress is tCO_2e per capita:

$$Total\ Emissions = Pop\ x\ CO_2/Pop$$
$$(Where\ CO_2/Pop = tCO_2e\ per\ capita)$$

Many countries have made commitments only to attempt to reduce their economic emission intensity ($kgCO_2e/\$-GDP$). As we just saw, this doesn't make any guarantee that emissions will reduce. So this commitment lacks any tangible emission reduction goal.

Next Kyoto Protocol targets – the Paris Agreement

The Kyoto Protocol emissions targets for Australia can be summarised as follows:

- Round one: Average emissions in 2008–2012 to be no more than 8% higher than in 1990.

- Round two: Average emissions in 2013–2020 to be 5% lower than in 2000.

- Round three: Emissions in 2030 to be 26-28% lower than in 2005. For Australia, this is now generally referred to as the Paris Agreement target.

Regarding the Kyoto Protocol round two target, there are a couple of factors in Australia's favour in meeting this target:

- As shown on the chart on page 41, 2000 was a high year for emissions with 548 $MtCO_2e$ emitted.

- Once again land use change sector emissions were a significant contributor, with 450,000 hectares of land clearing contributing to 66 MtCO2e emissions. So the new emission baseline is still high.

- As mentioned earlier, any improvement on the first target can be carried over to relax the second-round emission target. 'Carrying over' 5% means Australia already has a head start on its second-round target.

With an already minor emissions reduction in 2016 compared to the 2005 baseline, it looks like Australia will achieve the second 5% target for 2013–2020 – again without actually having to make significant fossil fuel emission reductions.

The third target, now generally known as the Paris Agreement target, represents a significant change in the degree of difficulty. The third target is proposed to be a 26–28% reduction by 2030, but this time using the baseline year of 2005.

Like the second target, this baseline year is favourable to Australia as it was a high emissions year, with:

- 589 $MtCO_2e$ emitted.

- land use change sector emissions again being a significant contributor, with around 500,000 hectares of land clearing contributing to 71 $MtCO_2e$ emissions.

- per capita fossil fuel emissions peaking in 2005 at 19 tCO_2e per capita.

However, that's where the advantages end. A 27% reduction on 2005 emissions means reducing them from 589 $MtCO_2e$ to 430 $MtCO_2e$. And with Australia's population expected to grow to around 29 million by 2030, the per capita emission would need to drop from 22 tCO_2e in 2016/7 to around 15 tCO_2e – a 33% reduction.

As mentioned earlier, Australia easily met both Kyoto emission targets by significant reduction in only one sector of the emission sector: land clearance in the land use change sector. Australia's relatively easy round one Kyoto target meant there was no long-term requirement to progress:

- tangible fossil fuel emission reduction initiatives (either in fuel and transport fuel efficiency improvements)

- a fuel mix change from coal to gas or greater levels of renewable energy.

For the Australian coal industry, Kyoto round one meant business as usual, and coal-fired generation capacity actually increased over the period.

The key question is, where will Australia's future emission reductions come from?

Land use change emissions are now calculated to be a net sink of CO_2 in Australia, with 24 $MtCO_2$ being reported as removed from the atmosphere in 2016 through reforestation. How much more the sector can achieve isn't yet clear, but it's unlikely it would reduce emissions enough to meet the 27% Paris target. To make further GHG reductions, emissions from other sectors will most likely need to be addressed, particularly fossil fuel consumption.

Here are the options available to reduce fossils fuel emissions.

1. Increase efficiency of traditional fossil fuel generation and consumption.

2. Replace high-emission intensity fuels with low-EI fuels (i.e. replace coal with gas).

3. Replace fossil fuels with renewable energy.

4. Reduce energy consumption.

5. Further outsource high-emission intensity industries such as oil refining, steel and aluminium manufacturing.

6. Perform carbon capture and storage (CCS) on tradition fossil fuel generation.

I'll discuss these options in more detail in Chapter 9.

Prior year adjustments to Land Use Change emissions

The 2015 NGGI report indicated that land use change net emissions were minus 7.7 $MtCO_2$, meaning a net absorption of CO_2 from the atmosphere from changes in vegetated land. The CO_2 absorbed by reforestation was calculated to be more than the CO_2 emitted from forest clearances.

In the 2016 NGGI report, this sink was revised up to -21.5 $MtCO_2$ due to an increase to the 2015 estimate of CO_2 absorbed by the effects of reforestation. A key question is whether this rate of reforestation CO_2 absorption can be further increased to contribute significantly to meeting the Paris emission reduction target.

This illustrates a key feature in the greenhouse gas international report obligations and methodology: emissions of prior years can be adjusted. This occurs for all sectors, but is significant predominantly in the land use change sector.

The following table shows adjustments for the years 1990 and 2015 made between the 2014, 2015 and 2016 NGGI reports.

NGGI Report year	2014	2016	2015	2016
Emission year	1990	1990	2015	2015
	$MtCO_2e$	$MtCO_2e$	$MtCO_2e$	$MtCO_2e$
Fossil fuel usage	256.8	256.8	375.0	376.2
Fugitive emissions	36.0	37.2	44.5	45.8
Industrial processes	26.1	26	32.3	33.8
Agriculture	80.1	80	70	70.1
Waste processing	19.7	20	11.4	11.7
Total excluding LULUCF	**418.7**	**420.0**	**533.2**	**537.6**
LULUCF	128.4	156.7	-7.7	-20.3
Total including LULUCF	**547.1**	**576.7**	**525.5**	**517.3**

Sources: [3], [4], [7]

It can be seen that:

- In 1990 (the Kyoto baseline year) land use change emissions have been increased by 28 MtCO2e.

- Along with other minor adjustments, this resulted in total emissions increasing by 30MtCO2e (1.8 tCO2e per capita).

- 2015 land use change emissions were reduced by 13 MtCO2e.

- Along with other minor adjustments, this resulted in total emissions decreasing by 8MtCO2e (0.3 tCO2e per capita).

The reason for prior year adjustments is to make necessary corrections (based on either new or previously missing information), or to account for new scientific methods of calculating an emission or its equivalence.

It isn't clear in the reports whether the Australian Government will use the 1990 re-adjusted emissions increases to reduce its Kyoto round two emission reduction obligations. (Remember, an improvement in the round one target achievement can be carried over as a reduction to the round two target.) Whether prior year adjustments should be used to change (i.e. make easier) an emission reduction target is an interesting point of debate. The fact a prior year emission is now estimated to be higher than originally thought means greater reductions are needed to compensate for the increase, regardless of whether the adjustment was calculated to be applicable in the baseline year. Whatever the emissions in previous years are adjusted to be doesn't change the actual emission and its effect on the atmosphere (if any).

It might be more appropriate to follow financial accounting procedures for carbon accounting. That is, once a year is closed out (from a financial reporting perspective) any later year adjustments must be made in the year they're found and reported, rather than going back and being able to change previously locked-down figures. This would have the effect of putting Australia's adjustment to 1990 land use change emissions in 2015/6, making it significantly harder to meet the Kyoto round two target.

CHAPTER 4

Australia's emissions in a global context

Here are estimates of carbon equivalent emissions for varies countries or country groups (in both absolute terms and per capita), based on their estimated population.

Country/Group	Pop. Mil	GHG MtCO$_2$e	% of Total	tCO$_2$e per capita
Australia	23.7	526	1.2%	22.2
Canada	36.7	738	1.7%	20.1
USA	326.2	6,280	14.4%	19.3
New Zealand	4.8	75	0.2%	15.5
Russia	146.8	2,199	5.0%	15.0
Germany	82.8	894	2.1%	10.8
Japan	126.7	1,353	3.1%	10.7
FSU [1]	142.8	1,284	2.9%	9.0
China	1,395.2	11,735	26.9%	8.4
United Kingdom	65.6	546	1.3%	8.3
FEB [2]	110.7	881	2.0%	8.0
EU-REST [3]	237.8	1,693	3.9%	7.1
Italy	60.5	421	1.0%	7.0

France	67.2	441	1.0%	6.6
OPEC Big 4 [4]	48.8	1,047	2.4%	21.5
OPEC Rest	440.4	2,136	4.9%	4.9
SEA [5]	723.4	2,726	6.3%	3.8
ROW [6]	2,082.0	5,696	13.1%	2.7
India	1,324.7	2,909	6.7%	2.2
TOTAL	**7,446.9**	**43,580**		**5.9**

Notes:
1. *Former Soviet Union (excluding Russia)*
2. *Former Eastern Block*
3. *Rest of Europe*
4. *Saudi Arabia, UAE, Kuwait, Qatar*
5. *South East Asia*
6. *Rest of World, including Africa (non-OPEC)*

Source: Various publicly available data sets [24], [25]. Collated and sorted into above groupings by the author.

The following key indicators can be inferred from the data:

- China has the highest total carbon equivalent emission of all country groups considered – around 12,000 $MtCO_2e/y$ or 27% of the world total

- Australia has the highest GHG per capita emission. However, Australia's absolute emissions are only about 1.2% of the world total.

- The UK has about the same total emissions as Australia. But as the UK has a far higher population, UK emissions on a per capita basis are only 34% that of Australia's.

- In absolute terms, India emits about 5.5 times as much CO_2e as Australia. However, due to the huge population difference the per capita emission for India is only 10% that of Australia.

So why are Australian emissions so much higher than the West's average?

To investigate this, the following section contains a comparison analysis of Australian and UK fuel and electricity consumption emissions.

Australia and UK consumption and emissions comparison

Comparisons of the carbon emissions of various countries aren't generally published in the GHG reporting requirements. In the International Panel on Climate Change (IPCC) reporting obligations, countries don't have to provide raw source data. However, we can gather raw consumption data from other reports.

The following carbon emission comparisons are derived from a combination of GHG emission reporting and fuel and energy consumption data for Australia and the UK. It uses a combination of 2015 and 2016 reports, and approximate conversion factors. UK emissions sector reporting has been mapped to the Australian sector reporting using assumptions shown in the UK raw data. It isn't perfectly consistent, but because the differences in emissions are so great, the analysis clearly demonstrates what drives the difference in the profiles of the two countries.

Country =>	UK	UK	UK	Australia
Population =>			65.6	23.7
Emission Sector	MtCO₂e 2015	%	tCO₂e per capita	tCO₂e per capita
Fossil fuel usage	396.1	80%	6.0	15.8
Fugitive emissions (1)	27.0	5%	0.4	1.9
Industrial processes	12.7	3%	0.2	1.4
Agriculture	49.1	10%	0.7	3
Waste processing	18.2	4%	0.3	0.5
Total excluding LULUCF	**503.1**	101%	7.7	**22.5**
LULUCF (2)	-7.4	-1%	-0.1	-0.3
Total including LULUCF	**495.7**	100%	**7.6**	**22.2**
		UK versus Australia =>	**34%**	

Notes:

1. *Methane emissions associated with fossil fuel extraction and use*
2. *Land Use, Land Use Change and Forestry*
Sources: [3], [11]

As you can see, the emissions from direct fossil fuel consumption (including fugitive emissions) are about the same percentage of total GHG emissions at 79% (Aus) and 84% (UK). However, UK absolute emissions (in 2015) were actually slightly lower than those of Australia's. And UK emissions on a per capita basis are only 34% of that of Australia's.

This is a dramatically lower emission performance. Let's investigate the possible drivers for this difference by considering raw energy consumption data for both countries.

Liquid petroleum fuel emissions

The UK Government reports liquid petroleum fuel consumption in a similar fashion to the Australian Government. The following table shows reported liquid petroleum fuel consumptions for the UK, along with the emissions factors estimated in Chapter 2.

Country =>	UK	UK	UK	Australia
Population=>			65.6	23.7
Fuel	ML 2016	L per capita	tCO$_2$ per cap	tCO$_2$ per cap
LPG	6,327	96	0.2	0.2
Petrol	18,110	276	0.6	1.8
Diesel fuels	32,134	490	1.3	2.6
Aviation fuel	14,249	217	0.5	0.9
Fuel oils	6,498	99	0.3	0.1
TOTAL	**77,318**	**1,178**	**2.9**	**5.7**
UK versus Australia =>		**51%**	**52%**	

Sources: [8], [12]

The results show that the per capita CO$_2$ emission from petroleum fuel usage in the UK is estimated at 2.94 tCO$_2$ – 52% of Australia's per capita emission. There isn't much difference in the liquid petroleum fuel mix emission intensity between the two countries. While the UK has a higher proportional usage of lower emission factor LPG, this is offset by a higher proportional usage of higher emission factor diesel and fuel oil.

The primary driver for lower emissions is simply that the UK consumes less fuel – around 1,200 litres per capita year versus 2,300 litres per capita for Australia.

Electricity generation emissions

This table shows reported electricity generation statistics, along with the emissions factors estimated in Chapter 2.

Country =>	UK	UK	UK	Australia
Population=>			65.6	23.7
Generation fuel	M-kWh 2016	%	tCO$_2$ per capita	tCO$_2$ per capita
Black coal	75,630	22%	1.0	3.9
Brown coal	0	0%	0.0	2.4
Gas	100,035	30%	0.6	0.9
Fuel oils	2,133	1%	0.0	0.2
Hydroelectricity	6,289	2%		
Wind (1)	47,872	14%		
Solar PV		0%		
Bio-generation	29,388	9%		
Nuclear	70,345	21%		
Geothermal	4,664	1%		
TOTAL	**336,356**	100%	**1.6**	**7.3**
kW per Capita	**0.59**			
UK versus Australia =>	**48%**		**22%**	

Notes:
1. *UK reports don't differentiate between wind and solar photovoltaic (PV) generation. For the UK it's assumed to be predominantly wind generation.*
Sources: [9], [12]

The per capita CO_2 emission from electricity generation and usage in the UK is estimated at 1.60 tCO_2, which is only 22% of Australia's per capita emission.

As with liquid fuel consumption, the table shows the UK simply consumes less electricity. It averages 0.59 kW per capita year versus 1.22 kW in Australia, which is about 48%. But unlike liquid fuel consumption, the emission intensity of the fuel generation mix is also a significant contributor to the lower UK emission. (I'll discuss this in more detail at the end of this chapter.)

Internal natural gas consumption emissions and oil refining fuel

This table shows reported gas and refining fuel usage statistics, along with estimated associated direct CO_2 emissions, for Australia and the UK.

Country =>	UK	UK	UK	Australia
Population=>			65.6	23.7
Natural gas - other uses	Mt	t per capita	tCO₂ per capita	tCO₂ per capita
Gas - all other use	35	0.5	1.4	2.6
Refining fuel usage	4	0.1	0.21	0.17
TOTAL	39	0.6	1.6	2.8
	UK versus Australia =>		59%	

Sources: [9], [12]

The per capita CO_2 emission from natural gas and refining fuel usage in the UK is estimated at 1.62 tCO_2 – around 60% of Australia's per capita emission. Again, this is primarily driven by lower per capita consumption (59% of Australia's consumption). Refinery fuel usage is the only area where the UK has a higher per capita consumption than Australia, and this is because the UK has a higher per capita oil refining capacity than Australia.

Australia has closed four of its eight large oil refineries over the past 15 years. However, the demand for refined products has actually increased over the same period to such an extent that Australia now imports around 55% of its refined petroleum product requirements.

Summary

Here's a summary of the CO_2 emissions (excluding fugitive emissions) just from fossil fuel consumption in Australia and the UK.

	Per capita consumption per year		tCO_2 per capita year	
Country =>	**UK**	**Australia**	**UK**	**Australia**
Liquid fuels - litres	1,179	2,307	2.9	5.7
Electricity - kW	0.59	1.22	1.6	7.3
Gas/refining fuel - tonnes	0.59	1.01	1.6	2.7
TOTAL			**6.2**	**15.7**
UK versus Australia =>			**39%**	

The lower emissions in the UK are due to two factors:

•   The UK's consumption of both liquid fuels and electricity is only about 50% of that in Australia.

•   UK fuel mix for electricity generation mix has only 45% of the emission intensity of that in Australia.

That's because:

•   The UK's black coal for power generation use is only 22% versus 63% in Australia

•   The UK has no brown coal power generation

•   The UK has 21% nuclear power generation versus nil in Australia

•   The UK has 14% renewable energy generation versus 6% in Australia

•   The UK has 9% bio-generation energy versus 2% in Australia.

So what *really* drives Australia to consume so much more fossil fuels than the UK? Influencing factors could include:

- Large land mass combined with low population density, resulting in longer transportation distances.

- Historical access to cheap and plentiful supplies of coal fossil fuel.

- Hotter climate driving a high reliance on refrigeration and air conditioning.

- Lower domestic natural gas reticulation and distribution. This means more electricity has to be used for direct heating, which is an inefficient use of electrical energy in terms of CO_2 emission.

- Natural resource availability, resulting in higher energy intensity economic/industrial activity, minerals extraction, and fossil fuel extraction.

Emissions relationship to a country's gross domestic product (GDP) and economic activities are discussed in Chapter 9. This will present a more detailed analysis of the factors driving the differences between not only the UK and Australia, but also other high- and low-emission economies.

CHAPTER 5

Energy pricing and the Levelised cost of Electricity (LCOE)

Y ou're probably familiar with the 'benchmark' crude oil price, which is often quoted in the popular media as being so many US dollars a barrel (US$/Bbl)[1]. Since 1990 this price per barrel has varied from around 13 US$ to as high as 140 US$. At the time of writing (2018) it's around 50 US$ (about 67 AU$) a barrel.

When a company buys crude oil or coal for industrial use, or a consumer buys petrol, electricity or natural gas for household use, energy is being traded. Unfortunately, the various energy products aren't traded in consistent units of measure.

Energy product	Pricing unit
Oil price	US$/Bbl
Petroleum liquids	$/L
Natural Gas	$/GJ (c/MJ for domestic retail gas supply)
Coal	$/tonne
Electricity	$/kWh or cents/kWh (c/kWh)

---

[1] Benchmark crude oil prices are always quoted in US dollars per barrel.

To make valid comparisons between the costs of the various energy types, we need to do some conversions. Sometimes it's useful to convert all the prices to $/kWh so they can be compared to the base unit cost of electrical energy. (The UK's domestic natural gas supply is priced in pence/kWh so it can be compared directly with the price of retail electricity.) However, the most appropriate units to use for a comparison often depend on the comparison being made, so it's worth taking a flexible approach to energy conversion.

The most useful conversion factors required are:

1 standard barrel of oil	=	159 L		
1 standard barrel of oil	=	6.1	GJ (gigajoules)	
1 GJ/tonne	=	1	MJ/kg (megajoules per kg)	
1 kWh	=	3.6	MJ =	0.0036 GJ

When considering the price of a raw energy product, the calorific (heat of combustion) value of the fuel is being measured. To estimate the cost of fuel when used to generate electricity, the price needs to be adjusted by how efficiently the fuel's calorific energy content converts into electrical energy.

This is similar to the formula used to estimate emission intensity in Chapter 2:

$/kWh (Electrical) = $/kWh (Calorific) / Efficiency

Here are examples of price comparisons we can make using the simple conversions I mentioned earlier.

Crude Oil Price Comparisons

	Crude Oil Price
US$/Bbl	50
AU$/US$	0.75
AU$/Bbl	66.67
$/L	0.419
GJ/Bbl	6.1
$/GJ	10.9
$/kWh (calorific)	0.022

Electricity generation fuel price comparisons

	Black coal	Brown coal [1]	Natural gas [2]	Diesel Retail [3]
Base Price $/t	75	n/a	n/a	1183
MJ/kg	24.00	10.00	52.50	44.00
$/GJ Price	3.13	1.50	10.00	26.90
c/kWh (calorific)	1.1	0.5	3.6	9.7
Efficiency	38%	31%	48%	44%
c/kWh (electrical)	3.0	1.7	7.5	22.0

Notes:
1. *Assumes brown coal cost is around 50% of black coal on a $/GJ basis.*
2. *Assumes an export natural gas price of 10 $/GJ.*
   (Historical gas price on Australia's east coast was 3–5 $/GJ pre-2015).
3. *Based on 100 c/L, and excludes excise duty (refundable on diesel purchased for generation use).*

As you can see, the lowest-cost fuel for generating electricity is brown coal (1.7 c/kWh) followed by black coal (3 c/kWh) and natural gas (7.5 c/kWh), with diesel (retail) being the highest (22 c/kWh).

However, this is just the fuel cost. The full cost of generating electricity has several other factors that need to be taken into account:

- Cost of financing the capital cost of the facility
- Fixed operating costs
- Variable operating cost (excluding fuel)
- Variable operating cost (fuel)

Analysing the full cost of electricity is known as Levelised Cost of Electricity (LCOE). I'll discuss this in the next section.

Levelised cost of electricity

The levelised cost of electricity (LCOE) is a measure of an electricity generation source that tries to compare different generation methods consistently. It is an economic assessment of the average total cost to build and operate a power-generating asset over its lifetime divided by the total energy output of the asset over that lifetime. The LCOE can also be regarded as the minimum cost electricity must be sold at to break even over the lifetime of the project.

LCOE methodology considers the cost of electricity to be made up of:

- Interest charges on the initial cost of capital to build the facility (finance costs)
- Fixed operating and maintenance costs over the life of the facility
- Variable operating and maintenance costs over the life of the facility
- Cost of fuel (nil for renewable energy)

These costs are estimated over the operating life of the facility. A discounted cash flow analysis is applied to discount future cash flows to current values, and the result is published as the LCOE in $/MWh[2].

A number of Government- and CSIRO-sponsored reports on LCOE have been written in recent years:

---

2 *100 $/MWh is equal to 10 c/kWh.*

- The Australian Energy Assessment 2013 Model update[17], produced by Office of the Chief Economist, Bureau of Resources and Energy Economics (BREE). Commonwealth of Australia.

- The Australian Power Generation Technology Report 2016 (APGT[13]), produced by a consortium led by CO2CRC consultants, with Steering Committee of CO2CRC, CSIRO, Australian Renewable Energy Agency, Department of Industry and Science – Office of the Chief Economist, Anlec R&D

- Independent Review into the Future Security of the National Electricity Market, Blueprint for the Future, Commonwealth of Australia 2017 (IRNEM[11]) by Dr Alan Finkel AO, Chief Scientist, Chair of the Expert Panel

As well as these official reports, there have also been a number of media-reported claims about the costs of electricity.

> *"But the Coalition campaign in favour of coal, and against renewables, is relentless, even after study after study is produced highlighting how wind and solar, even with integration costs, are a significantly cheaper option than sticking with a coal-based centralised grid In Australia"*
>
> **Renew Economy, March 2017**
> *https://reneweconomy.com.au/malcolm-turnbulls-trumpian-disregard-for-energy-facts-77393/*

> *"Whichever the case, let's be clear: clean coal is NOT A THING."*
>
> *"Large-scale wind and solar plants are already cheaper than new 'more efficient' coal plants."*
>
> **By The Climate Council, 02.02.2017**
> *https://www.climatecouncil.org.au/fact-check-turnbulls-speech-on-australias-energy-future*

> *"Some renewable power technologies are cost competitive with gas power generation now. Renewable energy sources such as solar PV and wind are already cost competitive, and prices are falling".*
>
> **Pollution and Price: The Cost of Investing in Gas[14], by The Climate Council Limited**

The first issue with these claims is that none of the official government or CSIRO reports actually support the idea that renewable energy is currently cheaper than fossil fuels on an LCOE basis. Both the APTG Report (2016) and the IRNEM Report (2017) show coal to be the cheapest form of electricity generation at the current time, see Appendix B.

What both reports *do* state is that renewables wind and solar are *forecast* to be comparable with the cost of coal-fired power generation sometime in the future.

To quote from page five of the APTG Report's executive summary:

> *"In the base case studied in this report, fossil-fuel technologies are the lowest cost generators, being lower than wind and significantly lower than solar PV."*

Here's a summary of the average LCOE costs in both reports.

Generating Source =>	Black Coal	Brown Coal	Gas	Solar PV	On-shore Wind
ATPG Report 2016					
2015 LCOE c/kWh	8.0	8.0	7.9 [(1)]	14.2	10.3
2030 LCOE c/kWh	7.5	7.8	9.0 [(2)]	8.0	7.0
IRNEM Report 2017					
2020 LCOE c/kWh	7.6	n/a	8.3	9.1	9.2
2030 LCOE c/kWh	7.5	n/a	9.3	6.1	7.9
2050 LCOE c/kWh	7.5	n/a	9.6	4.9	7.0

Notes:
1. *Using a current gas price of 6.5 $/GJ*
2. *Using a future gas price of 8 $/GJ*

The wind and solar PV generation costs are reduced because significant improvements in wind and solar PV implementation costs and performance are forecast for 2030.

The CSIRO's own website makes the following comments on the APTG Report, confirming that the forecast cost of renewable energy is predicted to be cheaper than fossil fuels:

> **Wind, solar, coal and gas to reach similar costs by 2030:**
>
> *By 2030 renewable energy sources such as solar and wind will cost a similar amount to fossils fuels such as coal and gas, thanks to falling technology costs, according to new forecasts released in the CO2CRC's Australian Power Generation Technology (APGT) Report.*
>
> *The convergence of conventional and renewable energy costs depends on the capital costs of these energy sources. These were modelled for the new report by CSIRO's Global and Local Learning Model. This model is a relatively objective way of projecting costs based on historical learning rates. Learning rates show that for each doubling of installed capacity of an energy source, costs fall by a particular amount. We can model these costs across different climate policies.*
>
> *https://blog.csiro.au/wind-solar-coal-gas-to-reach-similar-costs-by-2030/*

The APTG report also states:

> *The carbon price also has an impact, as it is initially only in the developed world but by 2020 it is in all regions of the model.*

This indicates that a carbon price (or tax) *has* been used as an input parameter to forecast renewable energy uptake, and therefore costs. A carbon price hasn't just been applied in Australia. It's been applied across the entire world by 2020, with the following outcome:

This means that more low-emissions technologies are needed to meet demand, which increases their cumulative capacity and thus cost reductions.

This might be considered to be a case of engineering a self-fulfilling prophecy. Because an increase in renewable energy demand has been legislated through a carbon price, its implementation cost will reduce, thus justifying the decision to legislate the incentive.

There may be another input assumption bias in the official LCOE studies. They both tacitly assume there will be downward pressure on coal consumption, either in Australia or worldwide, as part of a global drive to reduce emissions. While both reports consider forecast reductions in renewable energy costs in detail, neither one considers the possible effect of consumption restriction on the price of coal.

Renewable energy costs

So how and why can it be claimed that renewable energy is cheaper than fossil fuel energy? The LCOE calculation results are extremely sensitive to input assumptions (such as capital and operating cost assumptions), but most importantly for renewables the capacity factor.

The capacity factor is the average amount of power a facility produces as a factor of its maximum installed capacity. For fossil fuel generators, capacity factors are around 85%. Most of this 15% reduction is due to facility outages for maintenance periods. As maintenance periods can be scheduled with a reasonable degree of accuracy, periods of reduced production from fossil fuel facilities can be scheduled with reasonable degrees of certainty not to coincide with other facility outages, thus ensuring continuity of supply.

However, capacity factors are a bit more complicated for renewables. The reduction in power output due to the capacity factor is mainly caused by the variability of the power source (i.e. the strength of the wind and sunshine).

The average capacity factor for wind generation in the APGT study is 38%, and for solar PV it is 21%. This means a wind farm of 100 MW maximum capacity will produce electricity at an average rate of 38 MW over its lifetime. And a 100 MW solar PV facility will produce electricity at an average rate of 21 MW over its lifetime.

The AGTP report assumed a current capacity factor for wind power of 38%, rising to 45% by 2030. And for solar PV, the current capacity factor of 21% rises to 28% by 2030.

The problem is that although the official government-sponsored reports don't yet confirm that renewables are already cheaper than fossil fuels, plenty of other consulting organisations provide their own calculations and reports on the cost of electricity.

In its report *Pollution and Price: The Cost of Investing in Gas*[14], The Climate Council uses *Bloomberg New Energy Finance's New Energy Outlook 2016* as evidence that renewable energy is already cheaper than gas generation.

To quote from its report:

> *Some renewable power technologies are cost competitive with gas power generation now. As shown in Table 4, current gas power prices are already increasing due to increased gas costs. Renewable energy sources such as solar PV and wind are already cost competitive, and prices are falling. A range of other studies reinforce this point (e.g. Lazard 2016; Reputex 2017).*

The problem is, the report doesn't actually show that outcome.

- Electricity cost from natural gas is quoted as 62 US$/MWh, which is the lowest cost electricity in the evaluation.
- But the report assumes gas prices will rise to 14 $/GJ gas, increasing the cost of electricity from gas to 105 US$/MWh.

- Solar is quoted at 85 US$/MWh.

- Wind is quoted at 70 US$/MWh.

- Wind cost is stated as being as low as 55 US$/kWh for "some recent Australian wind contracts".

The cost of power from coal isn't mentioned in this report. But as it can be assumed from the other reports that coal is cheaper than non-export price gas, even this report seems to implicitly infer that coal is the cheapest form of electricity.

This report also uses improved cost and performance forecasts to predict future lower costs for renewables:

> *Wind and solar costs fall sharply. The levelised costs of generation per MWh for onshore wind will fall 41% by 2040, and solar photovoltaics by 60%, making these two technologies the cheapest ways of producing electricity in many countries during the 2020s and in most of the world in the 2030s.*
>
> *While already competitive in a number of countries today without policy support, the cost of onshore wind is expected to drop 41% by 2040, driven primarily by improving capacity factors – which reach 33% on average in 2030 and 41% in 2040. The solar experience curve also marches on, but decline in technology cost is increasingly accompanied by a reduction in the cost of development, finance and operation pushing new utility-scale solar down 60% from a $74-$220/MWh range today, to a central estimate of around $40/MWh worldwide in 2040.*

What about the claim that some recent wind contract prices are reported as low as 55 US$/MWh. The problem with using contract prices as a benchmark for the true cost of electricity from renewables is whether the price quoted includes adjustments for government subsidies provided for the facility's construction through renewable energy grants or other incentives.

To quote from the Australian Financial Review – Editorial 13/9/18

> *Victoria has now held a tender for renewables to help meet its 40% renewable energy target. Madly, in the auction the government has written "contracts for difference", which mean the governments pays the generator if prices are lower between the "strike price", and the generator pays the government if the price is higher: this makes the government a silent partner in renewable energy companies for 15 years. Victorian Labor claims this will help lower prices. But because the whole process has no transparency no one really has any idea how much the taxpayer is or could be on the hook for.*

Capacity factors and dispatchable supply

One problem with the capacity factors for wind and solar is that they are driven by weather conditions. But as the weather cannot be perfectly predicted, neither can the capacity factors. The capability to provide power on demand is called 'dispatchable supply'. The unpredictable capacity factors for renewable energy means it cannot be guaranteed to provide power 'on-demand', and is not considered dispatchable. In fact, the new acronym term for renewable energy is VRE – variable renewable energy.

We can illustrate the concept of dispatchable supply by considering the supply and demand profile for a typical household Solar PV installation:

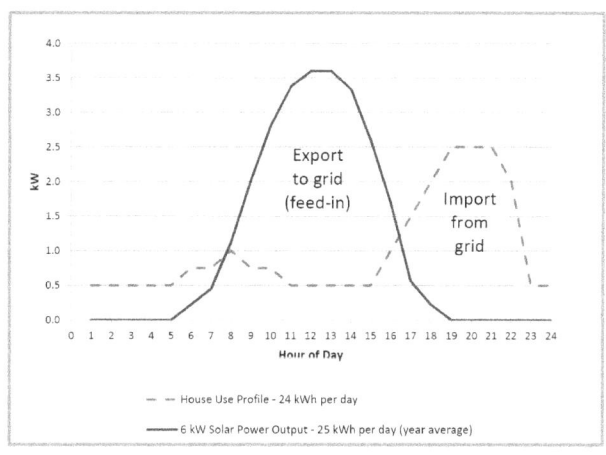

Source: [34]

Standalone solar power (i.e. without a battery) is only available when the sun is shining. And it varies over the course of a day as the sun rises and falls, as shown by the solid line in the chart. The dashed line shows a typical 'evening peak' household power usage profile.

A 6kW solar system in average conditions in Brisbane will output on average about 25 kWh per day (about 18% of the installed capacity, or 18% capacity factor). However, as the chart shows the power isn't necessarily available when the household needs it. The solar power only covers the end of the morning peak and the start of the evening peak. Without a battery storage system, the excess solar power is fed back into the grid. Electricity is sold back to the grid at an agreed price called the 'Feed-in tariff' (FIT). Example calculations for the economic returns on domestic scale solar PV and battery installations are given in Appendix A.

So while a solar unit can provide the average amount of electricity the house needs it can't meet the peak demand, and must be supplemented with other sources of dispatchable electricity.

Wind power doesn't follow the predictable availability of sunlight. But its electricity supply isn't dispatchable either as its availability can't be guaranteed.

At a utility grid level, electricity usage also shows a non-uniform demand pattern. The following chart shows electricity demand in Queensland over a one-week period (in January 2018) as a dashed line.

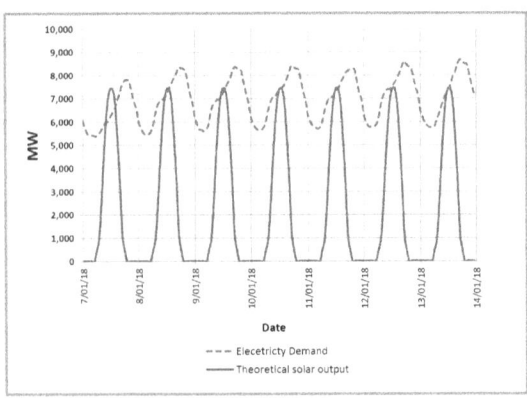

Source[30] *(Data for dashed line only, solid line added by author)*

A theoretical 7,000 MW of peak solar PV generation capacity has been overlayed onto this demand as a solid line. This illustrates the challenge of integrating so much renewable solar energy into the whole electricity system of Queensland. The 7,000 MW of solar power just about meets the average midday demand of 7,000 MW. However, that only covers about 31% of the State's total power demand (the difference in the areas under the two curves).

The individual household evening peaks sum up to an evening peak for the entire electricity system from about 6pm to 9pm (i.e. mostly around and after sunset). The solar power does not come close to meeting the evening peak demand of more than 8,000 MW.

To provide 100% of Queensland's power demand by solar, approximately 25,000 MW of peak solar power capacity would be needed. Only about 45% of the solar power would be used directly. Around 100,000 MWh of storage would be needed to provide power through to the start of solar production resuming the following morning. But that would only cover through to the next day. It would not provide any backup for prolonged dull weather.

To cover peak demand (or to guarantee to cover peak demand in the case of wind power), renewables need to be supplemented by one of the following:

- pump hydroelectric storage (PHES)
- large battery storage
- fossil fuel generation

The available data on domestic and large-scale batteries indicate they are still a very expensive way of providing energy storage. Add to that the life expectancy and end-of-life environmental disposal concerns, whether they will ever play a role in utility-scale storage for a zero net emissions economy is an interesting point of debate. The effect of battery storage on the cost of electricity from renewables wasn't considered in the official reports, but it was recommended that it be the subject of future studies.

The Snowy Mountain 2.0 project is a $4.5 billion tax payer funded 2,000 MW pumped hydroelectric storage proposal. It is intended to provide storage support for up to 3,500 MW of additional renewable generation production in NSW and Victoria[35]. Studies have been carried out to assess opportunities for further pumped hydroelectric storage in eastern Australia. Whether this has the potential to replace a significant portion of fossil fuel backup supply, and at what cost is the key question.

For now, the required backup generation capacity for variable renewable tends to be fossil fuel.

South Australia has driven significant growth in large-scale wind generation over the past few years. After suffering two serious supply interruptions during the 2016/7 summer, the SA government announced a $550m plan to resolve the state's energy issues in March 2017. This included instructing the state's network operators to install 200 MW of backup diesel generation capacity before the 2017/8 summer. The plan also included constructing a new 250 MW natural gas generation facility at a cost of around $360m, which was urgently progressed.

Supplementary fossil fuel generation has a very high levelised cost. Not because the facilities are more expensive to build than base-load generation facilities, but rather because of their lower utilisation. These facilities are used only for supplementary or backup demand when renewables can't produce it. So while the operator still incurs capital and fixed operating costs waiting to be called on to supply power, they're spread across fewer units of electricity sold. When they need to be used, the operators attempt to counter this by raising the price so they get as much income as possible from the short supply period to offset the periods where no electricity is sold.

However, these costs don't appear in the cost of electricity for wind generation.

The following comments about dispatchable supply are from the IRNEM Report.[1]

The increasing penetration of VRE (variable renewable energy) generators in the NEM (National Electricity Market) also has the potential to change how returns are distributed in the wholesale market. VRE generators, like wind and solar photovoltaic, are fundamentally different from traditional thermal generation assets in that they have no fuel cost and are non-dispatchable. Because these generators' short-run operating costs are essentially zero, they will generally bid into the wholesale market at zero, to ensure they are called to generate while other generators set the clearing price that is paid to all generators. At high penetration, VRE generators can become the price setter and set a zero or negative clearing price in the wholesale market.

At other times, the output from VRE generators can drop relatively suddenly (when there is thick cloud cover that affects solar photovoltaic or the wind stops blowing, affecting wind generators). This can require other generation to rapidly ramp up or down to balance system load – a capability for which many existing dispatchable generators are ill-suited. Rapid fluctuations in output also increases volatility in the wholesale spot market, potentially increasing the cost of managing price risk for electricity users.

The fact that VRE generators operate with no fuel cost could, with the right policy framework and with further technological development, be used to reduce overall wholesale prices. However, at present, VRE generators are increasing price volatility in the wholesale market and are creating challenging investment conditions for other generators.

What does this mean?

When the wind is blowing and the sun is shining, renewable generators undercut the market price and push their product into the market. Fossil fuel generators must lower their production to let renewable into the grid. For the fossil fuel generators, lower production means higher fixed operating costs per unit of electricity sold.

When the renewable generator's production drops due to lack of wind or cloud, the fossil fuel generators are called on to make up production. At this point they have no competition from the renewable generators, so they increase their price to make up for the revenue they lost during the renewable generator's operation. This is the increased *price volatility* noted above.

"Increasing the cost of managing price risk for electricity users" means the overall effect is the price of electricity is higher.

In my view market penetration of government-subsidised renewables into the mostly privatised electricity generation sector has completely thrown a spanner into the works of how the private sector is supposed to bring the cost benefits of free market competition to the consumer.

Conclusions - renewables

So what is the truth about the cost of variable renewable versus fossil fuel energy? As usual, the real truth probably lies somewhere between the extremes. My view is we don't have enough independent and impartial analysis to truly make a judgement on this at this time. But one thing appears clear. The simple levelized (or average) cost of electricity model analysis is not sophisticated enough to determine the true costs of integrating non-dispatchable renewables into a mostly fossil fuel network.

The problem isn't so much what the average price of electricity is, but what the incremental (or marginal) price of electricity is at peak demand, generally between 6 and 9pm in the evening. If renewables can't reliably meet peak demand, a dispatchable power supply needs to be on standby to make up the shortfall.

So here's another way to look at it: Who pays for the dispatchable generator to be on low rate or standby until it's needed? At the moment it's the privatised fossil fuel generators, who then try to pass that cost onto the consumer through higher wholesale electricity prices at peak demand. However some generators have determined that this is not providing sufficient economic return and have pulled out of providing the service. The two examples being the closure of the Hazelwood brown coal facility in Victoria and the announcement to close the Liddell coal fired power station in NSW.

But having said all that, the reports seem to indicate that renewables aren't actually that much different in costs to fossil fuels, and may well be cheaper in the future. The UK appears to have a more expensive power generation mix than Australia, with significant renewable and more expensive nuclear power. However, retail prices in the UK are lower than in Australia, demonstrating that higher generation costs can be adsorbed into the system without crashing the economy.

So a valid question is whether we as a society should absorb higher costs for the greater good of reduced emissions. After all, it isn't without precedent. As a society we phased out cheap lead additives in petrol for the improved health benefits. And the UK phased out its high sulphur coal fired power generation to reduce acid rain pollution problems in northern Europe.

My personal view is that we should be considering renewables in this context. That doesn't mean we ignore the costs. A relevant analysis of the true marginal costs of electricity generation options is essential to provide an informed debate.

To quote again from the IRNEM Report:

> The fact that VRE generators operate with no fuel cost could, with the right policy framework and with further technological development, be used to reduce overall wholesale prices. However, at present, VRE generators are increasing price volatility in the wholesale market and are creating challenging investment conditions for other generators.

I discuss the barriers to developing a coherent policy framework for energy and emission in Chapter 10.

Carbon capture and storage (CCS)

Carbon capture and storage (CCS) is a general term for processes that capture CO_2 from an emission source (such as a power station) and store the CO_2 so it doesn't enter the atmosphere.

The proposed storage is generally depleted underground fossil fuel reservoirs. Deep ocean storage has been considered, but there's concern it would exacerbate ocean acidification and is no longer considered a feasible option.

Injecting CO_2 recovered from natural gas processing facilities into depleted oil reservoirs is a proven technology. For decades it has been used for so-called enhanced oil recovery (EOR) from the reservoir. An estimated 50 $MtCO_2$ is injected in the USA per year, which is about 0.7% of its CO_2e emissions. EOR generates an economic return by extracting more oil and isn't specifically done to store the CO_2. Issues about the CO_2 potentially leaking from the reservoir back into the atmosphere aren't primary concerns in EOR projects. However, injecting CO_2 into the ground for CCS obviously requires a guarantee that the underground storage solution will never allow the CO_2 to leak back into the atmosphere, and this is a concern about the viability of the process.

While a number of different technologies are being researched, they all essentially follow these steps:

1. Recover CO_2 from exhaust flue gas, usually by adsorbing it in a solvent.

2. Recover the CO_2 from the solvent (which is then recycled).

3. Pump the CO_2 via a pipeline system.

4. Inject the CO_2 into the selected underground storage reservoir.

CCS itself is highly energy intensive. A CCS installed on a fossil fuel power station is estimated to consume between 25 and 40% of the power produced (sacrificial load). This effectively means we have to burn 25 to 40% more carbon to actually capture the carbon dioxide after combustion. This alone presents a risk to the viability of CCS. If any of selected reservoirs leak significant quantities of CO_2 back into the atmosphere, we will have generated more CO_2 trying to store it than would have been emitted through simple fossil fuel power generation.

CCS has been the subject of research and feasibility studies and appears to be technically feasible for some locations in Australia. The problem as always is the cost. The equipment needed to process thousands of litres of flue gas per second to recover the CO_2 (which is only a small percentage of the total flue gas composition) is a significant cost in itself. Then there's the cost of the pumping and the pipeline, along with the cost of injection.

The APGT Report[13] mentioned earlier indicates that the costs for CCS in Australia might range from as low as 5 $/tCO_2 to 70 $/tCO_2. Other references indicate that CCS will cost at least 60 USD/tCO_2. Although small-scale pilots have been constructed, no commercial-scale CCS facility solely for the sequestration of flue gas CO_2 (and not for EOR) has been put into operation. This means the costs for CCS are highly speculative. For illustration purposes, a cost of 100 $/tCO_2 (AU$) has been assumed in the examples in Chapter 2 and Chapter 9 of this book.

The APGT Report indicates that CCS will add around 12 c/kWh to the levelised cost of electricity (LCOE) from coal-fired power generation. It also states that a carbon price (or tax) of 130 $/tCO_2 is needed to make the price of electricity equivalent to a coal generator without CCS, which seem to confirm the 100 $/tCO_2 ballpark cost for CCS.

If CCS is the solution to worldwide carbon emissions, it's worth looking at the scale of the problem ahead. In the introduction I said every day the world consumes about 12 million tonnes of oil (around 85 million barrels), 21 million tonnes of coal, and 7.5 million tonnes of natural gas. Because of the 44/12 'expansion' of carbon into CO_2, this results in about 90 million tonnes of CO_2 emissions per day. The processing requirement of all the fossil-fuel-generated CO_2 (a theoretical scenario) would be 2.5 times the processing requirement of today's entire fossil fuel industry.

CHAPTER 6

The world in billions – Part 2:
The remaining carbon emission budget

"Let us eat and drink, for tomorrow we die."
—PAUL THE APOSTLE, CORINTHIANS 1 15:32

In Chapter 1 I said that human-induced carbon dioxide equivalent (CO_2e) greenhouse gas emissions for 2015 is estimated to be 44 $BtCO_2e$, with around 33 $BtCO_2$ attributed to fossil fuel consumption.

Humans have been burning fossil fuels for centuries. But it wasn't until the start of the modern industrial revolution (taken to be 1850) that fossil fuel consumption really started affecting the planet's natural (non-industrial activity) carbon balance. By 1900 emissions rose to an estimated 1.8 billion tCO_2 a year, but had risen to only 6.6 billion tCO_2 by 1950.

It's only over the past 50 years that fossil fuel consumption has really taken off. Since the beginning of industrial-scale emissions, human induced CO_2 emissions are estimated to have reached a cumulative volume of 2,000 $BtCO_2$ (2 trillion tCO_2).

Climate scientists have designed complex carbon cycle simulations to estimate how much total carbon the world can emit to stay within an acceptable (but not comfortable) temperature increase of no more than 2°C above pre-industrial levels.

While there's no exact answer, the estimate is somewhere between 2,600 and 3,600 BtCO₂ of total cumulative emissions[2]. We have already emitted an estimated 2,000 BtCO₂, which means we only have somewhere between 600 and 1,600 BtCO₂ to go – less than half of what we've burned so far.

The problem is we emitted that last 1,000 BtCO₂e (i.e. 50%) in just 40 years. Let's say we can safely emit another 1,000 BtCO₂e. Even if we gradually cut our current 44 BtCO₂e per year emissions to zero, we'll use up our remaining emission budget in about 50 years.

But in the same 40 years we emitted our last 1,000 BtCO₂e, the world's population has grown from 4.2 billion to 7.4 billion, with most of that growth (95%) occurring in the developing (Non-West) world.

Here's a copy of the chart from Chapter 1 with the estimated reduction we need to achieve.

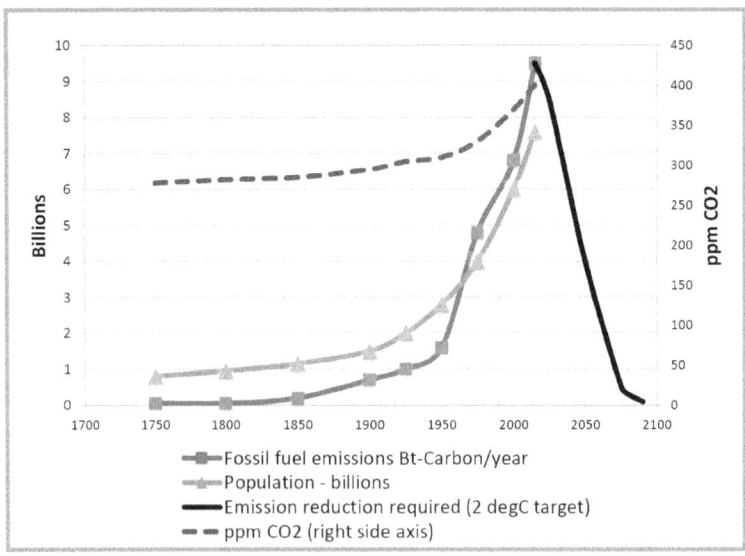

As you can see, to meet the 1,000 BtCO₂e limit we need to ramp down our global emissions about as fast as we recently ramped them up – starting today!

The problem with this goal is there are more fossil fuel reserves than the remaining carbon budget allows for. We have enough proven and probable fossil fuel reserves to emit an estimated 3,000 BtCO$_2$ into the atmosphere from today – three times the allowable budget. But if we include possible reserves and estimates of non-conventional fuel resources, that figure increases to a factor of about ten.

These countries have some of the largest estimated reserves of fossil fuels.

	Fossil Fuel Reserves, in BtCO$_2$ equivalent[2]
USA	513
Russia	450
China	245
Australia	165
Venezuela	137
Saudi Arabia	128

These big players alone have enough CO$_2$ emissions locked up in their reserves (1,638 BtCO$_2$) to bust our 1,000 BtCO$_2$ budget. The key question is, if the world is going to commit to limiting future CO$_2$ emission, how will it agree on the allocation of future consumption *and production* of fossil fuels? It's hard to imagine any world-governing organisation that could determine how to equitably split the world's remaining emission budget.

Carbon rationing

The fundamental problem the world faces could be considered how to allocate (or ration) the remaining carbon emission budget across the world's current and future populations.

To achieve this, we need to address three issues that, although they are relatively straightforward, will be extremely difficult to resolve:

1.   To cut emissions we need to cut fossil fuel consumption.

2.   To cut consumption we need to find a way to ration the remaining allowable consumption across the world's current and future population.

3.   *If* consumption can be rationed, we also need to ration fossil fuel production between the world's various energy production corporations, both private and government owned.

Unfortunately, here's no hope of an agreement on point one until we reach worldwide consensus and agreement on points two and three.

Regarding the second point: How could a current and future population's carbon ration be allocated? Perhaps we could give every citizen the same yearly carbon allowance from now on. The West would need to cut its consumption and emissions dramatically. However, the Non-West would argue even that isn't fair, as the West has already emitted far more carbon. So should cumulative carbon be taken into account in determining future allowances? If so, would it be calculated per capita of people now, or per capita of everyone who has lived since 1850?

As if resolving the second point isn't hard enough, resolving the third point will be an order of magnitude harder. Companies could extract and sell fossil fuels at a level far below the available reserves they have on their books. It would dramatically affect earnings and share prices, and the wealth of those holding interest in energy corporations. It's almost impossible to imagine the world's energy producing corporations being brought to an agreement on such an issue of production rationing.

This is the second part of the carbon collision course we are on.

CHAPTER 7

Australia's fossil fuel and minerals export economy

A ustralia is one of the largest exporters of fossil fuel coal and natural gas (as liquefied natural gas, or LNG) in the world. The international convention on greenhouse gas (GHG) emission reporting is for emissions to be accounted for in the country or administration where the fuel is consumed and the emissions arise. Fuel consumed while extracting and processing fuels and minerals for export is included in Australia's international GHG reporting obligations. However, there's no requirement to report on emissions arising from consuming exported fossil fuels in other countries.

It's worth looking at Australian fossil fuel exports in the context of their potential CO_2 emission. As shown earlier, natural gas electricity generation's CO_2 emission intensity (in $kgCO_2/kWh$) is about 40% that of a mix of black and brown coal. So substituting natural gas for coal-fired power generation would seemingly reduce Australia's reported GHG emission. However, Australia is one of the largest coal exporting nations in the world. If substituting natural gas for coal simply increases the availability of coal to be exported and consumed in other countries, the overall effect on world CO_2 emissions would be zero.

According to the September 2016 edition of *Resources and Energy Quarterly* (published by Department of Industry, Innovation and Science)[15], Australia's major fossil fuel exports for 2015–16 were:

Export commodity	Mt fuel (2016)	EF kg-CO_2/kg	$MtCO_2$
LNG	37	2.68	99
Metallurgical coal	188	2.20	414
Thermal black coal	200	2.20	440
TOTAL	**425**		**953**
	tCO₂ per Capita		**40**

Source [15]

Applying emission factors to these exports would result in a theoretical per capita CO_2 emission from fossil fuel export of 40 tCO_2 per capita year – almost double the internal Australian GHG emission given earlier (22 tCO_2e per capita year).

Indirect emissions – iron ore for steel manufacture

Australia largest mineral export by tonnage is iron ore at 786 million tonnes in 2016. Iron ore doesn't have a direct CO_2 emission when processed into steel, but it does have a large indirect emission. Iron ore is approximately 65% iron and 35% oxygen. To make steel, the oxygen is removed by heating the ore in a blast furnace with a source of carbon from metallurgical coal. The carbon reacts with the oxygen in the ore to form CO_2, leaving raw iron. Further processing is required to produce steel.

Large amounts of fossil fuel carbon and energy (from burning the metallurgical or coking coal) are needed to convert iron ore to steel. While it depends on the process, around 1.8 tonne of iron ore is needed to make one tonne of steel, with around 1.8 t of CO_2 being emitted in the process. So Australia's iron ore exports will indirectly result in about 786 million tCO_2e per year.

Note: Using exported Australian metallurgical coal for the iron ore processing would be double-counting the CO_2 emission. If all of Australia's metallurgical coal was used to process a portion of the Australian iron ore export, the CO_2 emissions from standalone Australian metallurgical coal would be deducted from this amount.

Indirect emissions – bauxite and alumina

Australia is one of the world's largest producers of aluminium ore (known as bauxite). Converting bauxite to aluminium is an energy intensive process, requiring around 14 kWh of electricity per kg of aluminium produced. Unless aluminium smelting is being done using only stranded hydroelectric power (power that would not otherwise be produced for general grid consumption), aluminium smelting is a very emission-intensive process.

Here's the process for converting bauxite to aluminium.

1. The bauxite is converted to pure aluminium oxide (alumina, Al_2O_3). This emits about 1.4 $kgCO_2$ per kg of aluminium in the alumina.

2. The alumina is converted to pure aluminium in an electrolytic smelter. This process itself emits about 1.8 $kgCO_2e$ per kg of aluminium produced, which is a combination of CO_2 from the removal of the oxygen in the alumina and other greenhouse gases emitted in the process.

However, most of the emissions from aluminium manufacture arise indirectly from the quantity of electricity needed to run the electrolytic smelter. The electro-chemical bond between aluminium and oxygen in alumina is extremely strong, and requires a large amount of electrical energy to break it.

Assuming the incremental electricity for aluminium production comes from coal-fired power generation at an emission intensity of 1 $kgCO_2/kWh$, the 14 kWh needed is equivalent to an indirect emission of about 14 $kgCO_2$ per kg of aluminium. This is about seven times the direct emission from the electrolysis smelter.

Australia produces and exports all three aluminium supply chain products: bauxite, alumina and aluminium. Here are the approximate quantities of these products, along with their inland/internal and exported estimated associated CO_2 emissions for 2016.

Product	Mt/y 2016	Al wt %	kgCO₂/ kg-Al	Mt CO₂ Australia	Mt CO₂ Exported
Bauxite - total mined	83.4				
Export grade [1]	19.2	23%	17.0		74
Inland grade [2]	64.2	18%			
Alumina production	21.8	53%	1.4	16.2	
Alumina export	18.8	53%	15.6		156
To Inland smelting	3.01	53%			
Aluminium production	1.60	100%	15.6	24.9	
To export	1.45				
Inland use	0.15				
TOTALS				**41.1**	**230**
t-CO₂ per Capita - Australia equivalent				**1.7**	**9.7**
Coal - International transportation requirement					93
Inerts - International transportation requirement					23.7

Notes:
1. *Bauxite - Export Grade is assumed to be 43% alumina.*
2. *Bauxite - Inland Grade is assumed to be 34% alumina.*
Source:
[15] data, analysis by author.
[32] 'The greenhouse gas intensity of Australian primary aluminium production, not including emissions from alumina refining which are considered separately, remained steady at 15.6 tonnes of CO2-e per tonne of aluminium'

This shows:

- Australia's aluminium production supply chain produces about 41 MtCO₂ (about 1.7 tCO₂ per capita), which is 7.6% of Australia's total emissions.

- Australia's aluminium supply chain material exports result in a further 230 MtCO₂ emitted by other countries.

Some people in the carbon emissions debate believe aluminium manufacturing/smelting in Australia should be stopped due to its high emissions. Because of those high emissions, a carbon price (tax) would significantly affect the cost of aluminium manufacturing in Australia. A 100 $/tCO$_2$ carbon price (tax) would increase the cost of Australia's aluminium by 1,730 $/t-Al, or 75%. It would almost certainly make aluminium smelting in Australia uncompetitive, and either significantly reduce or completely stop smelting in Australia.

The question is, would it actually reduce total global emissions? Or would the bauxite ore and alumina simply be processed in other countries, with global production and emissions staying the same?

World aluminium production – outsourcing emissions

World aluminium production is around 59 Mt/year. As you can see from this table showing aluminium production by country, China is the world's largest producer at about 32 Mt/year – 54% of the world's total production.

Rank 2016	Country	Mt/y 2016	% 2016	Emission reduction target
1	China	31.9	54%	No
2	Russia	3.6	6%	Revising
3	Canada	3.2	5%	Revising
4	India	2.9	5%	No
5	United Arab Emirates	2.5	4%	No
6	Australia	1.6	3%	Yes
7	Norway	1.2	2%	Yes
8	Bahrain	1.0	2%	No
9	Saudi Arabia	0.9	2%	No
10	United States	0.8	1%	Withdrawn
	Rest of World	9.2	16%	
	Total	58.8		

Source: [26]

It's worth noting that eight of the world's top ten aluminium smelting countries either don't have emission reduction target commitments, have withdrawn from some of them, or are revising their commitment[33].

If stopping production in Australia just increases production in other countries, it does nothing to reduce global carbon emissions from aluminium smelting. It simply outsources the emissions to elsewhere in the world.

In fact, it may actually increase overall emissions. Bauxite and alumina contain only around 40% and 53% aluminium respectively. To export them as raw materials, a large volume of inert waste by-products also needs to be transported. Around 24 Mt of the 38 Mt of the bauxite and alumina exported by Australia are waste by-products.

So what would happen if Australia converted all of its aluminium raw materials to aluminium 'in-house'? While Australia's internally reported CO_2 emissions would increase by 230 MTCO2, international transportation requirements for waste by-products would decrease by some 24 Mt/year.

But this isn't counting any coal that has to be transported. If the electricity for aluminium manufacture is being produced from coal, up to around 90 Mt of coal would be required. International transportation requirements could be even higher.

Reducing the fuel and waste material international transportation requirements from producing more aluminium in Australia may actually *reduce* overall global carbon emissions.

This simplified analysis illustrates the point that emissions must be considered in an overall context. Shutting down smelting industries in some countries because they have high emission footprints won't necessarily result in an overall reduction in global emissions. It may simply move the emissions elsewhere, and may even result in higher global emissions due to increased transportation requirements.

World steel production 1990-2016

World steel production is estimated to be around 1,620 Mt/year. Again, the world's largest producer is China with around 808 Mt/year – about 50% of the world's total production.

	Country	Mt 1990	Mt 2016	2016 versus 1990	Emission reduction target
Rank	World	770.4	1,620.4	850.0	
1	China	66.4	808.4	742.0	No
2	Japan	110.3	104.8	-5.5	Revising
3	India	15.0	95.6	80.6	No
4	United States	89.7	78.6	-11.1	Withdrawn
5	Russia	55.0	70.8	15.8	Revising
6	South Korea	23.1	68.6	45.5	Yes
7	Germany	44.0	42.1	-1.9	Yes
8	Turkey	9.4	33.2	23.8	No (Kyoto)
9	Brazil	20.6	30.2	9.6	Yes
10	Ukraine	54.6	24.2	-30.4	Revising
15	Spain	12.9	13.6	0.7	Yes
16	Canada	12.3	12.7	0.4	Revising
19	United Kingdom	17.8	7.6	-10.2	Yes
25	Australia	6.6	5.3	-1.3	Yes
38	UAE	0.1	3.2	3.1	No
	Others	232.6	221.5	-11.1	

Source: [27]

Similar to aluminium, most of the top steel producing countries either don't have emission reduction target commitments, have withdrawn from them, or are revising their commitment.

Overall steel production increased by about 850 Mt/year (+100%) between 1990 and 2016. The table shows the UK had the highest percentage decline in steel production in that time. It's also classified as a contributor to the UK's reduced GHG emissions during the same period.

To quote the UK's 2015 GHG Emission Report[11]:

> *In 2015, UK emissions of the basket of seven greenhouse gases covered by the Kyoto Protocol were estimated to be 495.7 million tonnes carbon dioxide equivalent (MtCO2e), a decrease of 3.8 per cent compared to the 2014 figure of 515.1 million tonnes.*
>
> *This decrease in emissions was mainly caused by:*
>
> - *A decrease of 2.6 per cent (2.3 MtCO2e) in the business sector, driven by a reduction in emissions from fuel used in the iron and steel sector due to the closure of one of the UK's three integrated steelworks in September 2015*

Although the steel production league table had a few ups and downs, almost the entire net steel production increase of 850 Mt/year has been in China (742 Mt/year) and India (81 Mt/year). The UK's production hasn't been "lost" to the world. It has simply been outsourced to other (and generally cheaper) jurisdictions.

As I mentioned earlier, international convention on emissions reporting requires jurisdictions to report emissions arising only from their own inland consumption. The UK reduced its CO_2e emissions by more than 30% between 1990 and 2010. However, it also relies more and more on imported goods. It has been estimated that if CO_2 emissions from the balance of imported and exported goods are taken into account, UK emissions actually rose by an estimated 20% during this period[2]. These emissions, generally from heavy industrial activity, have been outsourced in many cases to countries that have less onerous emission reduction commitments.

Similarly, Australia has reduced its CO_2 emissions over recent years by reducing oil refining capacity. In the past 20 years, four of Australia's eight large-scale oil refineries have been shut down, reducing its oil refining capacity by about 400,000 barrels per day – about half of Australia's petroleum product requirement. Refined fuel consumption has increased by about 7% over the past five years, resulting in a net import requirement increase of about 140% (about 30,000 million litres per year, 55% of Australia's refined product consumption of around 55,000 million litres per year). This production has been outsourced to refineries in countries such as Korea, Singapore, Japan, India, China and Malaysia. Australia has reduced its reported CO_2 emission by about 0.2 tCO_2 per capita by importing refined product instead of producing it 'in-house'. However, the emission hasn't disappeared. It has simply been outsourced to other countries.

Australia's major resource exports 1990 – 2015 (Kyoto period)

As mentioned earlier, Australia's easy round one Kyoto target meant there was no long-term requirement to progress:

- tangible carbon emission reduction initiatives
- fuel or transport fuel efficiency improvements
- generation fuel change from coal to gas
- greater levels of renewable energy.

For the internal consumption of Australian coal, Kyoto Stage one meant business as usual.

Internal coal consumption for power generation during 1990–2016 is shown in the following chart:

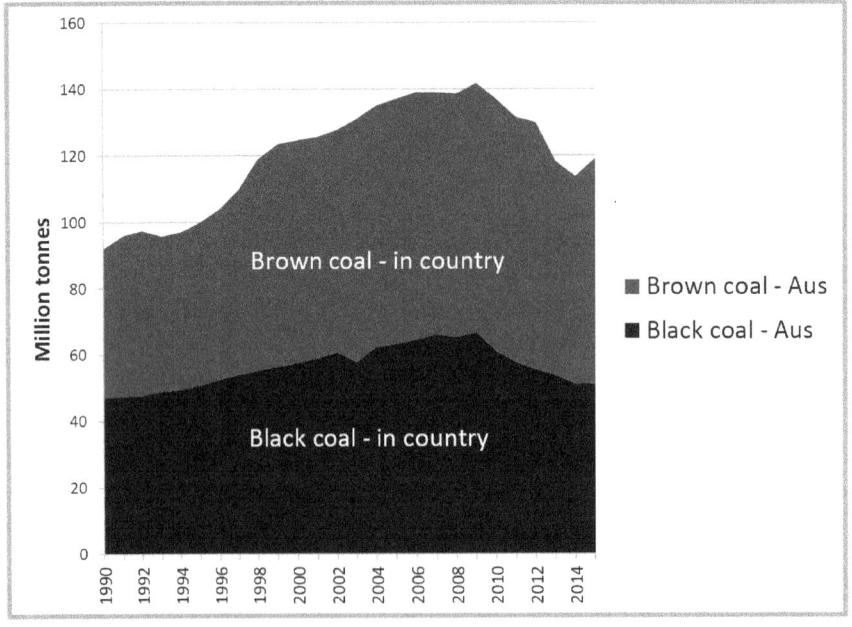

Sources: [9], [15]

Despite recently declining after peaking in 2010, Australia's overall internal coal consumption has increased from about 90 Mt/year to 118 Mt/year.

However, the recent decline in internal coal use doesn't mean Australian coal production is declining. Here's the same internal coal consumption as the previous chart, but this time including coal exports.

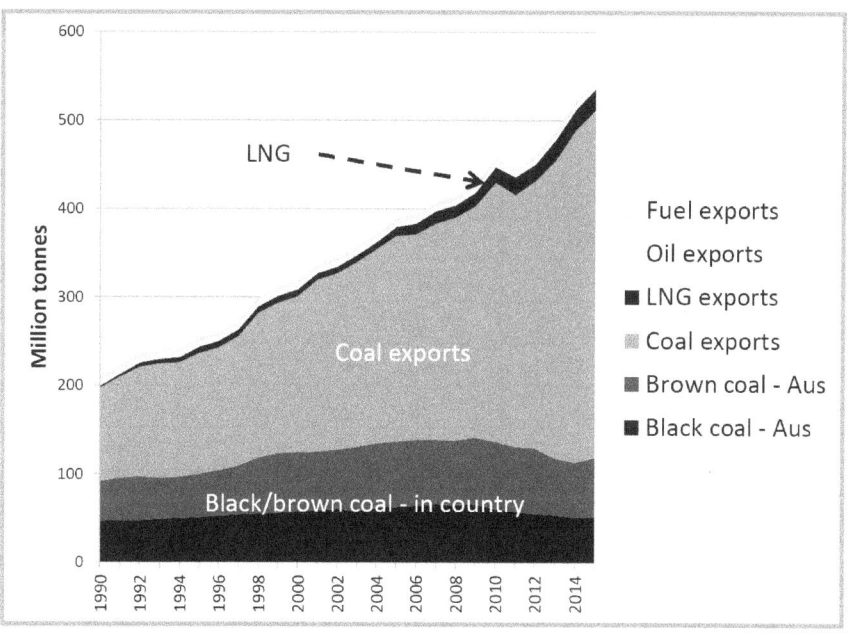

Sources: [9], [15]

Australia's total coal production has grown from 200 Mt in 1990 to more than 500 Mt, with exports rising from around 100 Mt to 390 Mt.

Here's a summary of Australia's major resource exports over the period 1990–2015.

Commodity	1990	2015/6	2015/6 vs 1990
Volume Mt			
LNG	3.4	36.9	1080%
Coal	113.6	389.3	340%
Petroleum	10.5	13.1	120%
Total Fossil Fuels Mt	**127**	**439**	**340%**
Iron Ore	104.0	785.8	760%
Bauxite	7.0	21.0	300%
Alumina	9.2	17.7	190%
Aluminium	0.9	1.4	160%
Total Ores Mt	**121**	**826**	**680%**
Revenues M$			
Fossil Fuel Only	**9,957**	**57,469**	580%
Ores Only	**7,466**	**58,027**	780%
All Resources	**28,674**	**160,376**	560%
GDP M$	759,586	1,617,933	210%
% GDP all Resources	**3.8%**	**9.9%**	

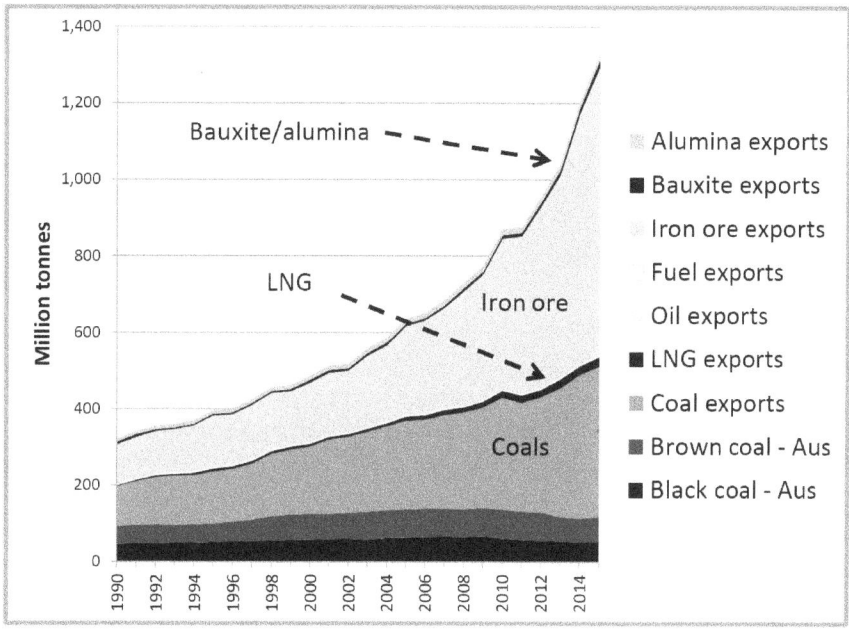

Sources: [9], [15]

To summarise:

- Australia's direct emission fossil fuel exports have increased by a factor of 3.4.

- Non-direct emission mineral exports have increased by a factor of about six.

- Revenues from all mineral resource exports have increased by a factor of six.

In 2015, fossil fuel exports contributed $57 billion to the Australian economy. In the same year, all resource exports (gold, metals, iron ore, bauxite, alumina and uranium, etc.) contributed around $160 billion. It has grown from around 4% of GDP in 1990 to around 10% of GDP as Australia reduced its other heavy industrial and carbon-intense activities such as car manufacturing, steel and aluminium smelting, and oil refining.

Here's the key issue: If Australia transitions completely to a zero-emission economy (which would mitigate global carbon emissions by only 1.2%), what should happen to its resource exports such as:

- those with direct carbon emissions (fossil fuels)?

- those with indirect high carbon emissions intensities, such as iron ore, bauxite and alumina?

If coal is phased out for Australian power generation, should all coal, gas and petroleum exploration also be phased out?

The Climate Council of Australia certainly thinks so. From their report *Energy Communications Guide 2018*[16]:

> *For a 75% chance of limiting global temperature rise to less than 2°C, more than two-thirds of known global gas reserves cannot be burned. Based on an economic analysis of how much coal, oil and gas can be burned for the world to have a 50% probability of keeping global temperature rise to 2°C and meeting our commitments under the Paris Agreement, 95% of Australia's coal reserves and 51% of our gas reserves must be left in the ground (McGlade and Ekins 2015; Climate Council 2017e). These reserves are unburnable. This means that in order to be consistent with our pledge to join the global effort to limit global temperature rise to less than 2°C, exploration of new reserves of conventional gas should end and there should be no development of unconventional gas. In order to tackle climate change, Australia also cannot bring online any new coal (Climate Council 2015).*

In the previous chapter it was shown that Australia's fossil fuel reserves are equivalent to about 165 $BtCO_2$. Assuming this is predominantly low-value coal at an emission factor of 2 $kgCO_2/kg$ and a price of 50 $/t, that's about $4,000 billion in potential revenue. Would it be politically palatable and electable for an Australian government to propose such a write-off? Even if it won an election with such a policy, what's to stop it being repealed by a future government the same way the carbon tax was repealed in 2014?

CHAPTER 8

Carbon offsets –
are they too good to be true?

Many high-consumption fossil fuel businesses (e.g. airlines and power retailers) give their customers the chance to offset the carbon emissions associated with consuming their products, usually for a small fee.

On the face of it, it sounds like an opportunity to reduce emissions. However, the scientific and environmental community is sceptical about whether these carbon offset schemes achieve any meaningful long-term removal of CO_2 from the atmosphere.

Another concern is the businesses offering these schemes are encouraging more consumption by having consumers believe the environmental harm has been mitigated.

The carbon offset schemes generally fall into these broad categories:

- Tree planting/reforestation
- Renewable energy projects
- Energy efficiency improvements
- Investment in collecting and neutralising refrigerant and similar GHGs

- Investment in projects to reduce existing emissions of GHGs

- Payments made to hold off clearing forest otherwise earmarked for deforestation.

The primary concerns with these offset schemes are:

- Do they genuinely remove the quantity of carbon they claim they do?

- Is that capture permanent and irreversible?

Most standards for CO_2 offsets require a guarantee that it be captured for 100 years. But it's almost impossible to see how it can be guaranteed and verified. Even if it *can* be verified, it will likely take centuries for our carbon emission to be mitigated to the point where our atmospheric CO_2 levels are back to pre-industrial levels. Even 100 years doesn't seem long enough to ensure the CO_2 doesn't reappear.

Here's a summary of concerns regarding the validity of these offsetting methods.

Tree planting/reforestation

It's hard to argue that planting trees isn't a good thing. But how permanent is the carbon capture? Trees are susceptible to burning, disease, death and eventual decay, which releases CO_2 back into the atmosphere. (See the carbon cycle in Chapter 2). Whether tree-planting schemes can capture carbon long enough to significantly affect global CO_2 levels is something we don't have enough operating experience to verify. What's more, we can't determine exactly how much CO_2 trees and forestation projects can capture. It can only be estimated.

Renewable energy projects

Whether these truly lead to 'additional' emission reductions is a controversial topic. Renewable energy reduces carbon emission only if it displaces an equivalent amount of conventionally produced electricity from the local grid. A significant amount of renewable energy brought online in the past 50 years has been used to supplement conventional energy. In other words, the total energy demand has gone up at a higher rate than renewable energy has come online.

Even if the renewable energy *does* displace an equivalent amount of conventionally produced electricity from the local grid, it won't necessarily reduce global emissions. Reducing coal consumption for electricity generation in Australia makes more (and cheaper) coal available for export. Unless the fossil fuel coal it would have burnt is left permanently in the ground there's no 'additional' emission reduction.

Energy efficiency improvements

The problem here is similar to that of renewable energy. Do efficiency improvements really lead to 'additional' emission reductions? Historically, every efficiency improvement we've made in the past 100 years has allowed us to consume more energy for the same cost – *not* the same amount of energy at a reduced cost. (This is known as the **Rebound effect**[2].)

These types of reductions are also susceptible to the **Export diversion effect**. For example, having Australia transition from coal to natural gas won't reduce emissions if the displaced coal is exported. Again, unless the fossil fuel coal that would have been burnt is left permanently in the ground there's no 'additional' emission reduction.

Collecting and neutralising refrigerant and similar GHGs Projects to reduce existing emissions of GHGs

These initiatives are all subject to **Perverse Incentive** – the incentive to increase emissions to claim a higher reduction credit when the emission offset is sold. This became prevalent in the HFC industry, and offsets are now no longer available for HFC-23 destruction.

In the oil and gas industry, companies have claimed offset funding for projects to combust (flare) raw methane that would otherwise be vented directly, reducing the overall CO_2e emission. The question is, shouldn't this be regulated anyway, rather than using the offset to let another entity emit the equivalent amount of CO_2?

Payments made to hold off clearing forest otherwise earmarked de-forestation

Here's a paragraph from an airline's offset scheme website:

> *Conserving Tasmania's Wilderness. Tasmania is home to one of the world's last great temperate wilderness areas and is renowned for its unique species, biological diversity and the country's largest tract of temperate rainforest. This carbon offset project protects over 7,000 hectares of native Tasmanian forest which would, if not for the project, continue to undergo selective logging or be cleared and converted to pasture.*

The questions here are similar to those on the previous issue:

- Shouldn't this wilderness be protected anyway?
- Has the logging and clearing requirement gone away, or has it just been moved to another location?

Apart from these concerns, carbon offset schemes are remarkably cheap. But are they too good to be true?

Airline offset example

The following text is taken from a major airline's carbon offset website:

> We understand the impact air travel has on the environment, and that's why we're committed to finding ways to help make a difference.
>
> Since 2007 our scheme, with your help, has offset over 3 million tonnes of carbon emissions, making us the largest offsetter of any airline.
>
> Air travel is a fast and convenient way for us to cover long distances but the planes we fly on release greenhouse gas emissions, which are shown to have negative effects on our natural environment.
>
> We've calculated the fuel emissions for each route we fly so you can easily offset your share, helping to counter these effects by supporting projects that affect real change.
>
> 100% of your contribution goes towards verified carbon offset projects that meet strict international standards, including the Verified Carbon Standard and the Gold Standard. These projects collectively help mitigate environmental impact and nourish communities.
>
> 1. Calculate... Calculate the emissions impact that your flights will generate.
>
> 2. Contemplate... Choose to protect forests, support developing communities or renewable energy.
>
> 3. Celebrate... Feel good about cleaning the climate and creating a sustainable future.

This seems to affirm the potential marketing intention of the service. Rather than feel guilty or concerned about resulting emissions from using the service, consumers should celebrate the consumption as their emission has been offset.

How realistic is it for a relatively small fee to offset the entire CO_2 emission of an airline flight?

Using the airline's own web-based calculator, the emission from a 17,000 km trip from Sydney to London is calculated to be 1.975 tCO_2 per passenger. The cost to offset this amount of CO_2 is $26, or 13 $/$tCO_2$. Using an estimated ticket price of $800, the CO_2 emissions are offset for an additional fee of around 3%.

That seems remarkably cost effective, considering various references on the potential for carbon mitigation by carbon capture and storage (CCS) technology anticipate the costs to be *at least* 100 $/$tCO_2$.

Does this seem too good to be true? And is it a worthwhile contribution to mitigating global carbon emissions?

Firstly, if the CO_2 emitted is indeed 100% mitigated, why not just put the 3% surcharge on the price of all fuel to offset all the CO_2 emission?

Secondly, the airline claims to offset 450 tCO_2 per day, making it "the largest offsetter of any airline". That's equivalent to 0.164 million t CO_2/year. That may sound significant, but the airline Environmental Performance Report (available online) shows the company's aviation fuel consumption results in emissions of about 11.7 million t CO_2/ year. So even the largest airline offsetter can claim it offsets only around 1.4% of its total CO_2 emission.

Thirdly, the online offset calculator takes only fuel CO_2 emission into account. Vapour trails from aircraft are estimated to double the effective CO_2e (equivalent) emission over the base fuel CO_2 emission[2].

The offset calculator doesn't provide an estimate for business-class travel either. As discussed in Chapter 2, one web-based calculator estimates that international business class travel emits 3.4 times as much CO_2 per passenger as economy class.

Power company example – tree planting

One Australian electricity retailer is currently (2017/8) allowing consumers to offset all of the carbon dioxide emission from generating the electricity used in their house at no additional costs.

To offset the carbon emissions associated with your household electricity you would have to plant around 25 trees each year. Or you can take a few minutes now to go carbon neutral. It costs you nothing to do something good.

We will offset 100% of the carbon produced by your home's electricity use at absolutely no cost to you. Here's how it works:

We help you to make your electricity at home carbon neutral by purchasing carbon offset units from a range of Australian and international offset projects which can include:

- *renewable energy projects in developing countries*
- *land management*
- *tree planting in Australia and more*
- *It's simple, effective and certified*

Our carbon neutral program has been certified against the Australian Government's National Carbon Offset Standard to provide carbon neutral electricity. To offset the amount of carbon released into the atmosphere we buy carbon offset units from a range of projects across the globe and right here in Australia.

Our data shows that the average Australian household consumes about five megawatt hours (MWh) of electricity per year and the average emissions intensity of that electricity is around one tonne of carbon dioxide equivalent (tCO2e) per MWh. On this basis, the average Australian household's electricity usage would account for five t CO2-e each year.

There is no single formula to calculate exactly how much carbon is absorbed by trees over their lifetime. It varies depending on the species and age of the tree, region it is grown, rainfall, and many other factors.

Based on a Technical Report from the South Australian Department of Environment Water and Natural Resources[1], Trees For Life Pty Inc. calculates that five trees will sequester around one t CO_2-e over thirty years, based on their tree plantings in South Australia at a density of 1000 trees per hectare at an annual carbon sequestration rate of 6.67 CO_2 per hectare.

Source: https://www.treesforlife.org.au/carbon/carbon-facts

Therefore, you would need to plant around 25 trees each year and maintain them for at least 30 years to offset an average household's annual electricity emissions.

Or you could opt to Go Neutral today...

... at no cost to our customers

Does this also seem too good to be true? If an electricity retailer can offset 100% of CO_2 emissions solely by tree planting schemes at no costs to the customer, why do we even have a carbon emission problem? Why can't we just apply this low-cost solution across the entire electricity generation system and offset everyone's carbon emissions at zero cost?

Sydney Opera House – carbon neutral

The Sydney Opera House recently declared itself to be certified as a 100% carbon neutral business. It achieved this not by eliminating CO_2 emissions arising from its operations, but by purchasing international carbon offsets.

The data on the Opera House's carbon neutral program is available online. Its annual carbon footprint is calculated to be 17,598 tCO_2e per year. Eighty-four percent (14,747 tCO_2 per year) of this emission is from electricity consumption. Assuming an average emission intensity of 0.7 kgCO_2/kWh, on average the Opera House consumes about 21,000 MWh of electricity, an average continuous consumption of 2,400 kW – the equivalent of about 2,400 family homes.

The Opera House purchased most of its carbon offset units from 105 MW of proposed solar power generation projects to be installed by the Siam Solar Energy 1 Co Ltd in Thailand. The facilities are expected to generate 149 MWh of electricity per year. (This is equivalent to an average power production of 17 MW, which means the facility has an average capacity factor of about 16%.)

The carbon offset seller's reports assume that:

1. All the electricity produced by the new solar power generation facility will exactly replace power that would otherwise be generated by local fossil fuel generators at an emission intensity of 0.54 kg CO_2/kWh. This will result in a carbon offset claim for the whole project of about 80,000 tCO_2 per year.

2. None of the fossil fuel that would have been consumed to generate this power will be sold on for consumption elsewhere (i.e. it will be left in the ground).

3. Proceeding with the project without the sales of carbon offset credits would have been uneconomic.

    "The project developer has demonstrated that the proposed project in the absence of any external financial support would not have been implemented. The report proves that the project would not be economically or financially feasible without the revenue from the sale of Gold Standard Voluntary Emission Reductions (VERs)." Source[29].

The primary concern with using such an offset to claim carbon neutrality is ensuring the fossil fuel claimed to be backed out isn't just extracted and consumed somewhere else. Thailand is a relatively poor country, with a per capita gross domestic product of around $6,000 (USD). Is it realistic to expect that any fossil fuel reserve available for economic extraction won't be extracted and used elsewhere in the region?

It's easy to demonstrate that the Sydney Opera House is subsidising solar power generation in Thailand.

It's also easy to demonstrate that the amount of solar power generated would produce a calculated quantity of CO_2 if that power was generated by conventional fossil fuel.

However, it's almost impossible to prove the fossil fuel that would have been consumed won't be extracted and consumed as fuel elsewhere in the region. In a world where every unit of fossil fuel energy that can be economically extracted and sold *is* extracted and sold, any claim of carbon neutrality cannot actually be proven.

It might be more accurate to describe it as a charitable endeavour to subsidise solar power generation in the developing world, rather than a true global carbon neutral scheme.

Carbon offsets – conclusions

Are carbon offsets and neutrality schemes doing any good at all? As I said earlier, it's the subject of some debate. Some environmental campaigners believe they raise awareness and achieve some good environmental protection measure, and so are a good thing.

But offsets have also been accused of fostering an 'indulgence mentality' in consumers, where they indulge in their high emission activities even more because they don't think they're contributing to the emission problem.

To quote from the New York Times[1]:

> **Carbon-Neutral Is Hip, but Is It Green?**
>
> *By ANDREW C. REVKINAPRIL 29, 2007*
>
> *"The worst of the carbon-offset programs resemble the Catholic Church's sale of indulgences back before the Reformation," said Denis Hayes, the president of the Bullitt Foundation, an*

---

[1] *http://www.nytimes.com/2007/04/29/weekinreview/29revkin.html*

environmental grant-making group. "Instead of reducing their carbon footprints, people take private jets and stretch limos, and then think they can buy an indulgence to forgive their sins."

Also from the New York Times[2]:

Paying More for Flights Eases Guilt, Not Emissions

By ELISABETH ROSENTHALNOV. 17, 2009

In 2002 Responsible Travel became one of the first travel companies to offer customers the option of buying so-called carbon offsets to counter the planet-warming emissions generated by their airline flights. But last month Responsible Travel canceled the program, saying that while it might help travelers feel virtuous, it was not helping to reduce global emissions. In fact, company officials said, it might even encourage some people to travel or consume more.

"The carbon offset has become this magic pill, a kind of get-out-of-jail-free card," Justin Francis, the managing director of Responsible Travel, one of the world's largest green travel companies to embrace environmental sustainability, said in an interview. "It's seductive to the consumer who says, 'It's $4 and I'm carbon-neutral, so I can fly all I want.'"

Of course, it could never be argued that tree planting, reforestation and protecting natural environments are bad things. But do the claims made by offsetting schemes that we can offset emissions at little if any cost really measure up? If they did, we wouldn't have an emissions problem at all.

We can't prove or disprove whether carbon offsetting has any measurable effect on atmospheric CO_2 levels. But if the world's larger airline offsetter can only offset 1.4% of its total emissions, it's unlikely to significantly reduce global CO_2 levels.

---

[2] *http://www.nytimes.com/2009/11/18/science/earth/18offset.html*

Unfortunately, my conclusion is that carbon offsets are a modern-day 'false prophet' of emissions mitigation. We can't escape the real inconvenient truth that to cut emissions we need to cut consumption. Offsetting may encourage some to think we can just carry on as normal, which is a dangerous precedent if we really want to do something about reducing our fossil fuel consumption emissions.

CHAPTER 9

The world in billions – Part 3: Carbon emissions and gross domestic product

"And again I say unto you, it is easier for a camel to go through the eye of a needle than for a rich man to enter into the Kingdom of God."
–ATTRIBUTED TO JESUS CHRIST – MATTHEW 19.24

Chapter 1 ('The world in billions – Part 1') described CO_2 emission per capita on a global level, and showed that on average the world emits about 6 tCO_2 per capita. This chapter looks at per capita emissions in relation to economic wealth, measured by Gross domestic Product (GDP) in US$.

The chart on the following page shows the estimated population (horizontal axis) and GDP per capita[28] of various countries and country groups first shown in Chapter 4. The area under each rectangle is proportional to the value of the economic output of each country or country group.

The chart shows how the population of the world can be split into two clearly defined groups.

The "West" consists entirely of the western industrial economies of North America, Western Europe, Australia, New Zealand and Japan, and has about one billion people. Those living in the West have high wealth and high consumption, particularly of fossil fuel. The average GDP in the West is about US$ 41,000 per capita.

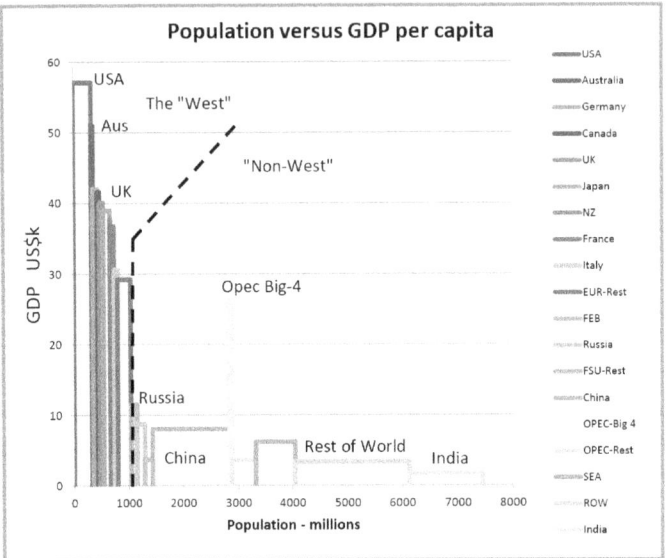

By contrast, 6.4 billion people living in developing economies (the "Non-West") have an average GDP of only US$ 4,700 per capita.

What's interesting to note is the stark division line between the two groups.

- Almost every group in the West has an average GDP of more than US$ 30,000 per capita.

- Almost every group in the Non-West has an average GDP of less than US$ 10,000 per capita.

It is a close approximation of the 'Pareto Principal' (after Vilfredo Pareto 1848 – 1923), which is 80% of the wealth is owned by only 20% of the population. (This is also commonly known as the 80-20 rule). In this case 14% of the world population living in the West has around 60% of the world's wealth.

Here's the same chart, but with CO₂e emissions added.

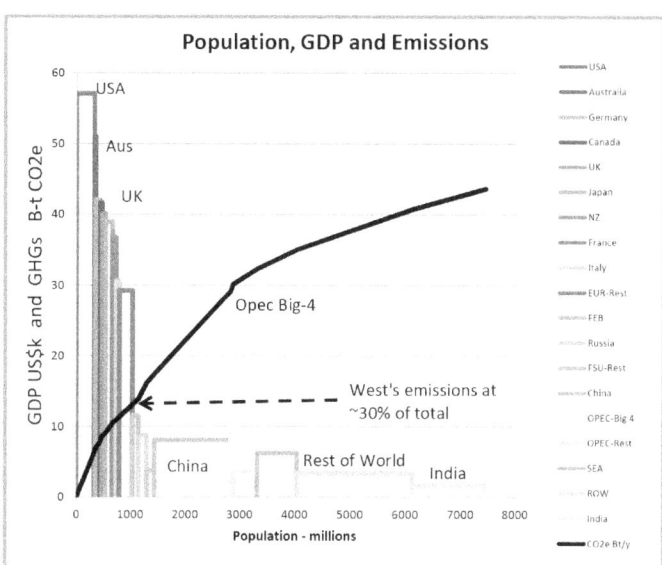

The solid line shows the cumulative CO2e emission contribution from each country or group, summing to the world's total of about 44 billion tonnes CO2e per year.

This chart indicates a correlation between wealth and emissions.

The slope, or steepness of the emissions line is proportional to the per capita emission for each country or group. So the wealthiest western countries (USA, Australia, Germany and Canada) lead with the highest slope on the curve. Apart from increases in steepness at Russia, China and the Opec Big-4 countries, the slope of the curve generally reduces as it moves right through the less affluent countries and groups.

Here's a summary of the differences between the West and the Non-West.

	West	Non-West	Total
Population - billions	1.0	6.4	7.4
Wealth - US$ trillions	44	31	75
% of Wealth	59%	41%	
GDP per capita US$,000	42.5	4.8	10.1
Emissions Bt-CO_2e	13.0	30.6	43.6
% of Emissions	30%	70%	
tCO_2e per capita	12.6	4.8	5.9

This division between the two groups for emissions is less stark than for wealth. But the West still emits on average about 2.6 times more CO_2e per capita than the Non-West.

Wealthier nations don't consume more energy than poorer ones just because they're wealthy. Having access to cheap fossil fuel energy (both historically and currently) has allowed the West to build the wealth building infrastructure it enjoys today.

The only country group in the Non-West with comparable wealth to the West is the big energy country group, the OPEC Big 4. And it has emission levels to go with it.

Looking at these figures, it's easy to see the dilemma we face.

While the 6.4 billion people in the Non-Western economies emit only about 5 tCO_2e per capita, collectively they now emit 70% of the world's CO_2e emissions. What's more, they now want their fair share of fossil fuel consumption to build their wealth as the West has done in the past century.

Even if the West diverted a portion of its wealth into completely mitigating carbon emissions (e.g. a combination of 100% renewable energy and carbon capture and storage (CCS)), it would be completely negated by the 6.4 billion people in the Non-West increasing their emissions by only 2.0 tCO_2e per capita.

Another potentially disturbing scenario is population growth. The world's population is expected to increase by two billion over the next 20 to 30 years, again mostly in developing nations. Even if consumption remains at 4.8 tCO₂e per capita, this growth in population would also negate any CO₂e mitigation the West might be able to achieve.

It's little wonder that the world's combined scientific and political might hasn't had much (if any) success in reducing carbon emissions.

Emissions versus GDP

This chart shows per capita GDP versus per capita carbon emissions for the same countries or groups.

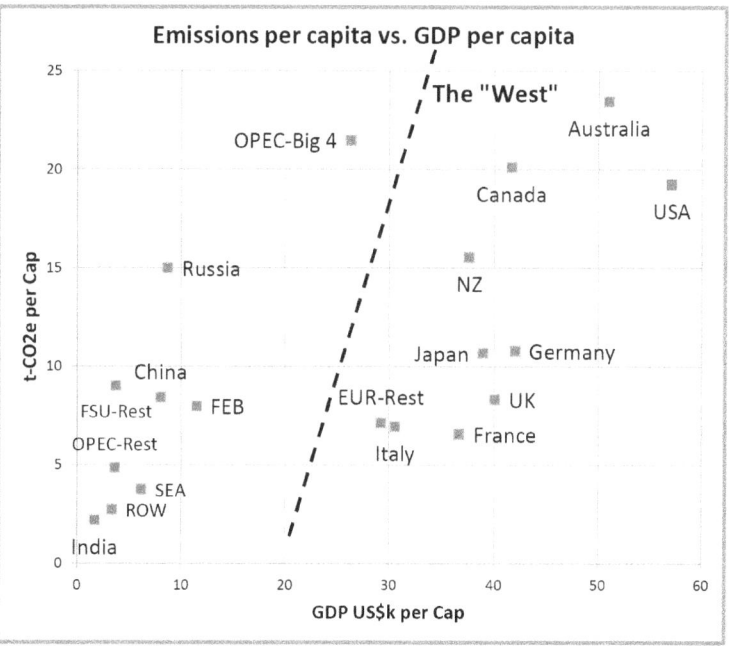

Again, the West sits at the high end of the both the emissions and GDP scales. And there's still a correlation between wealth and emissions. But there's also a considerable variation in the levels of emission within the West, ranging from 7 to 8 tCO₂e per capita in Italy and the UK to 19, 20 and 22 tCO₂e per capita in the USA, Canada and Australia respectively.

Five countries/groups have emission running at high levels in relation to GDP:

- Australia
- USA
- Canada
- OPEC Big 4 (Saudi Arabia, Kuwait, UAE, Qatar)
- Russia

These countries/country groups all have:

1. High fossil fuel/mineral resource extraction economies, which in itself leads to high energy demand for the industry of fuel/minerals extraction

2. Access to locally sourced, cheap fossil fuels for internal use, which reduces the incentives for energy efficiency and renewable energy development within the country

3. High levels of metal smelting industries, especially aluminium. Russia, Canada, UAE, Saudi Arabia, USA and Australia are all in the world's top 10 aluminium smelting countries.

And many of the countries (particularly Australia) have large land areas with low population density, driving internal transportation and distribution fuel consumptions requirements.

This suggests that high natural resources and fossil fuel extraction activity is in itself a key driver for high energy consumption, and therefore emissions. Access to cheap energy also may also reduce incentives for energy conservation measures such as smaller, more efficient cars. This may be the key reason why these economies have emissions so much higher than the other Western nations.

The following table shows a list of countries with emission levels greater than 20 tCO$_2$e per capita.

Country	Pop. Mil	t CO_2e per capita	GDP US$ per capita
Kuwait	4.1	47.4	26.6
Brunei	0.4	45.6	26.8
Qatar	2.6	31.4	59.5
Belize	0.4	25.5	4.6
United Arab Emirates	9.4	23.6	39.5
Bahrain	1.5	23.0	22.0
Oman	4.6	22.3	13.8
Australia	23.7	22.1	52.0
Libya	6.4	20.9	5.2
Canada	36.7	20.1	41.6

It shows almost all of the countries with per capita emissions greater than 20 tCO2e are primarily large fossil fuel extraction economies. With the exception of Belize and Lybia, all of the non-western countries are relatively wealthy oil exporting nations.

Most of the countries with high emission to GDP ratios either never agreed to the Kyoto or Paris Agreement targets, have withdrawn, or are redefining their proposed commitment.

USA: Never ratified Kyoto Protocol, are in process of withdrawing from the Paris Agreement

Opec Big 4: Not part of Kyoto Protocol. No reduction commitments under the Paris Agreement

Canada: Withdrew from Kytoto Protocol. Now part of Paris Agreement

Russia: Part of Paris Agreement, but redefining actual commitment

The charts show it is possible to run reasonable wealthy economies at less than 12 tCO2e per capita of emissions. However, these countries either have lower natural resources and minerals economies or have already used much of them up, as is the case for the UK. They have more diverse industrial manufacturing base activities (Germany, France and Japan) and/or a high services industry sector (UK).

The UK lower emissions (compared to Australia) are due to a number of factors:

1. Its transition from coal to natural gas electricity generation, though not intentionally. The UK ran out of cheap coal in the 1980s, but found an abundance of natural gas in the North Sea. That natural gas is now running low, and the UK now draws gas all the way across Europe from Russia.

2. The UK has high cost but low emissions nuclear power generation facilities (21%). This certainly wouldn't have been intended solely for emission reduction, but for security of supply and weapons purposes.

3. The UK is quite windy. It now generates around 14% of its electricity from wind power, and is further developing significant off-shore wind facilities.

4. The UK has moved away from heavy industries such as steel manufacturing. To quote its 2015 GHG Emission Report: "A decrease of 2.6 per cent (2.3 $MtCO_2e$) in the business sector, driven by a reduction in emissions from fuel used in the iron and steel sector due to the closure of one of the UK's three integrated steelworks in September 2015". (See Chapter 7.)

5. The UK is a small country a with a high population density, which means its fuel requirement for transportation (both private and business) is lower than Australia's.

6. The UK has historically placed very high excise taxes on petrol and diesel, encouraging people to take up high fuel efficiency/lower emission vehicles.

The key question for Australia is whether it should be chasing lower emissions by transition its economy away from energy intensive activity. It might be very difficult to do now, given the degree the economy has moved towards specialisation in the fossil fuel extraction and export industry over the past 30 years.

Potential to reduce Australia's fossil fuel emissions

As shown in Chapter 3, Australia easily met both Kyoto emission targets even though it made significant reductions in only one emission sector: land clearance/land use change (LULUCF sector). Australia's relatively easy round one Kyoto target meant there was no long-term need to progress tangible fossil fuel emission reduction initiatives through either:

- fuel and transport fuel efficiency improvements
- a fuel mix change from coal to gas or greater levels of renewable energy.

Land use change emissions are now calculated to be a net sink of CO_2 in Australia, with 24 $MtCO_2$ reportedly removed from the atmosphere in 2016 through reforestation. While it isn't yet clear how much more the sector can achieve, it's unlikely it would reduce emissions enough to meet the 27% Paris target requirements. To make further GHG reductions, emissions from other sectors will likely need to be addressed, particularly fossil fuel consumption.

Here's a summary of the options available to reduce fossils fuel emissions (which I highlighted in Chapter 3):

1. Increase efficiency of tradition fossil fuel generation and consumption.
2. Replace high-emission-intensity fuels with low-emission-intensity fuels (i.e. replace coal with natural gas).
3. Replace fossil fuels in general with renewable energy.

4.   Reduce energy consumption and wastage.

5.   Further outsource high-emission-intensity industries such as oil refining and steel and aluminium manufacturing.

6.   Perform Carbon Capture and Storage (CCS) on tradition fossil fuel generation.

Increase efficiency of tradition fossil fuel generation and consumption

The laws of thermodynamics limit the efficiency of fossil fuel generation to about 50–55%. We've been developing fossil fuel generation for the past 150 years, and it has just about reached its ceiling. High-efficiency combined-cycle gas turbines power stations (CCGT) can already achieve efficiencies of about 48% in Australian conditions (3–5% higher in colder European condition).

Coal-fired generation efficiency is about 38% (black coal) and 32% (brown coal), not because of poor design or lack of development but rather because coal (especially brown coal) has a high moisture content. Much of its inefficiency is due to having to dry the coal as part of the combustion process, which consumes energy. High-efficiency/low-emission (HELE) coal generation (discussed in Chapter 2) can improve coal generation efficiency, but with only a 10–12% improvement in emissions.

So efficiency improvements alone are unlikely to present a significant opportunity to reduce emissions.

Replace high-emission-intensity fuels with low-emission-intensity fuels

Generating 159,000 million kWh of electricity with lower emission intensity natural gas instead of coal-fired power could halve emissions from 170 MtCO2e to 85 MtCO2e – a reduction of about 3.7 tCO2 per capita. In today's terms that would reduce total emissions by about 16%. Not enough to hit the Paris Agreement target, but still a fair improvement.

However, a switch to natural gas to reduce emissions is not without its complexities.

- What happens to the coal that's displaced? If it stays in the ground, then it really would reduce emissions. But if it's exported to another country and burned there, there's no net reduction in emission. (See Chapter 7 on Fossil Fuel Exports for more details.)

- In its report *Pollution and Price: The Cost of Investing in Gas*[14], The Climate Council suggests that gas is not sufficiently less polluting than coal to provide any overall emission reduction benefit. The issue is the volume of fugitive (escaping) methane associated with fossil fuel extraction:

  - How much methane greenhouse gas escapes to the atmosphere during extraction and processing of gas compared to coal? (Both result in emissions of methane, coal also results in other greenhouse effects from nitrates, sulphur and particulates).

  - What is the true global warming potential (GWP) of methane? Historically, methane was assumed to have a GWP of 25 x CO_2 (on a 100 year basis). Some research[36] proposes a higher figure of 90 should be used, which a 20 year basis.

Resolution of these highly complex issues is required before a decision to switch to natural gas for emissions reduction reasons can be taken.

Gas prices in Australia have risen significantly over the past five years, from historical levels of around 4 $/GJ to 10-12 $/GJ. (I discuss the drivers affecting the price of gas on Australia's east coast in the next chapter.)

At 12 $/GJ, wholesale electricity from gas-fired power generation is about 4.5 c/kWh more expensive than from than coal-fired generation. Replacing 159,000 million kWh of coal-fired power generation with gas would cost about $300 per capita year. And that doesn't take into account the retail uplift on the higher wholesale price.

Replace fossil fuels with renewable energy

Renewable energy was discussed in detail in Chapter 5. The main issues with renewable energy arising in that chapter are:

- What are the true costs?
- Should the costs be calculated on an average basis or on a marginal peak demand basis?
- Who pays for the backup and/or storage capacity that is needed to meet peak demand?

But as I also said in Chapter 5, the reports seem to indicate that renewables aren't actually that much different in costs to fossil fuels. The UK appears to have a more expensive power generation mix than Australia, with significant renewable and more expensive nuclear power. However, retail prices in the UK are lower than in Australia, demonstrating that higher generation costs can be adsorbed into the system without crashing the economy.

So a valid question is whether we as a society should absorb potentially higher costs for the greater good of reduced emissions.

As I also said in Chapter 5, my personal view is that we should be considering renewables in this context. That doesn't mean we ignore the costs. A relevant analysis of the true marginal costs of electricity generation options is essential to provide an informed debate.

Bio-mass fuel

In recent years the UK has invested in (or more correctly subsidised) another arena of renewable energy: bio-mass fuel. The UK has converted two-thirds of one of its largest coal-fired power stations (the 1,900 MW Drax facility in Yorkshire) to run entirely on wood pellets. The final third is currently slated for conversion soon, as the UK plans to phase out all coal-fired generation by 2025. Approximately seven million tonnes of wood pellets are imported from the USA and Canada to run the power station.

The wood pellets are reported to be made from by-products of the wood harvesting industry such as offcuts and sawmill dust.

The calculations of the resulting overall carbon footprint for acquiring, transporting and burning the wood pellets aren't publicly available. But like many issues in the renewable energy debate, they are the subject of considerable disagreement.

This disagreement is essentially around whether the wood product is genuinely being harvested as a true carbon-neutral product over the long-term life of the forest. One side claims it is, while the other side claims it is not proven to be carbon neutral and that net emission are actually higher than using fossil fuel[31].

The cost of transportation has been suggested as a possible impediment to bio-mass take up in Australia[1]. But *if* the UK can successfully transport bio-mass fuel across the Atlantic with a low to no net carbon footprint, surely bio-mass fuel could be an option for Australia. Using bio-mass fuels in conventional power generation facilities does have the advantage of providing dispatchable, or on-demand power.

It would be beneficial to investigate how much carbon neutral the UK's bio-wood fuel scheme really is, and whether a similar scheme could successfully be applied in Australia.

Reduce energy consumption and wastage

Lowering emissions by reducing the amount of energy consumed can fall into two broad categories:

- **Consume less.** This could mean not travelling as far (by car or aircraft) and using less electricity in the home by using less air conditioning or watching less TV.

- **Use energy more effectively (or waste less energy).** This could mean travelling the same distance as before, but in a more fuel-efficient vehicle. This isn't conversion efficiency, but rather more *effective* use of energy. Using a heavy vehicle when a light vehicle could do the same job wastes energy moving the extra weight of the vehicle.

It could also include other energy conservation measures, such as improving building insulation standards.

As we saw in Chapter 4, the UK's petrol and diesel consumption is only about 50% of Australia's. Here's the price of petrol in Australia compared to six of our Western industrial peers (USA, UK, and Europe's Big 4: France, Germany, Italy and Netherlands).

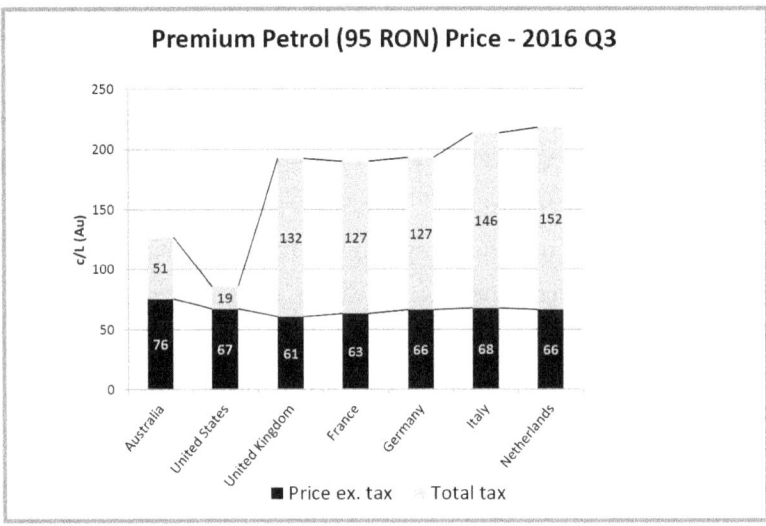

Source: [8]

Out of all these countries, Australia has the second-lowest petrol price. (The USA has the lowest price.) And there's a reasonable correlation between petrol prices and emissions, with the USA and Australia having the cheapest retail petrol price and correspondingly high emissions.

The price difference is driven by tax. Without it, Australia's price is around 10 c/L more than all the other countries.

Duties and taxes in the UK add 132 c/L to the pre-tax price, making the final price around 65 c/L more expensive than in Australia. Even though the UK liquid fuels excise duty isn't called a carbon tax, it effectively is one. The petrol price difference between the UK and Australia is equivalent to an incremental carbon tax of around 270 $/tCO_2.

Registration cost and other taxes in the UK also encourage:

- High uptake of smaller, higher efficiency, lower emission vehicles
- High uptake of diesel in light passenger vehicles
- High update of manual transmission vehicles, which are around 10% more fuel efficient than those with automatic transmissions.

Should Australia follow this approach, and increase taxes on liquid fuels to reign in consumption? It could be done. But given the often hysterical reporting of even minor petrol price increases in the media, it's hard to imagine the Australian electorate being persuaded to increase taxes on petrol consumption of even 23.5 c/L (the 100 $/tCO_2 carbon tax equivalent).

Other usage improvements worth considering are better regulations on energy wastage, particularly in buildings. For example, Australian shopping malls are often open to the outside, allowing the cool air-conditioned air to escape into the environment. Tighter regulations on this type of waste would surely improve the effectiveness of our energy consumption.

Outsource high-emission-intensity industries

Outsourcing high-emission-intensity industries has already reduced Australia's reported emissions. During the past 15 years it has closed four of its eight large oil refineries. However, in the same period Australia's demand for refined products has increased to the point that it now imports around 55% of its refined product requirements. The emission from crude oil refining operations has effectively been outsourced.

There are two problems with this approach:

- Moving the same emission-intense activity to another country doesn't reduce global emissions.

- It raises concerns around jobs and the economy as a whole.

It's also a politically touchy topic. It's one thing for certain sectors in the debate to say the West doesn't need heavy industry anymore, and that we should transition the economy to focus on new technologies and services sectors. But it's another thing to determine whether it's in the best economic interests of current and future populations.

As I talked out in Chapter 7, outsourcing heavy industry has helped the UK reduce its emissions during the Kyoto Protocol years. But if we take the net carbon footprint on the balance of imports and exports into consideration, UK emissions are estimated to have increased by more than 20%.

It's also worth noting that Australia's wealth in the world rankings (using per capita GDP) is second only to that of the USA. Having plentiful supplies of fossil fuels for its heavy mineral extraction industry almost certainly factors into Australia being one of the wealthiest countries in the world.

Perform Carbon Capture and Storage (CCS)

Carbon capture and storage (CCS) for fossil fuel power generation has been discussed for a number of years. However, it's still unproven on a commercial scale and is predicted to be expensive. The estimated cost of CCS in Australia is around 100 $/tCO$_2$ – the equivalent of adding about 10 c/kWh to the cost of electricity from coal-fired power generation.

Summary

It seems unlikely that the measures discussed in the previous section would allow Australia to realistically transition to a zero-emission economy without affecting the cost and convenience of energy use in Australia. But when compared to the UK, Australia can almost certainly do more to lower emissions by more efficient and effective use of energy. The key question is whether the electorate will vote for greater restriction on the freedom of choice in energy use they currently enjoy.

Theoretical impact of mitigating worldwide carbon emissions by carbon capture and storage

Carbon capture and storage (CCS) has been espoused as the process by which the world's carbon emissions could be mitigated.

The estimated cost of CCS for electricity generation is around 60 USD/tCO$_2$, which may well be optimistically low. No fully-functioning CCS demonstration plant has been successfully designed and put into operation, so the cost is highly speculative.

However, we can add that theoretical cost to the previous section's GDP data to estimate the costs of carbon mitigation by CCS as a percentage of GDP:

		West	Non-West	China	India
Carbon Emission	Bt-CO$_2$e	13.5	31.5	11.7	2.9
Cost of CCS	US$/t-CO$_2$	60	60	60	60
Total Cost	B-US$	810	1,890	704	175
Population	Billion	1.000	6.500	1.395	1.300
Cost per Cap	US$/person	810	291	505	134
GDP per Cap	US$/person	42,500	4,800	11,232	1,700
% of GDP		1.9%	6.1%	4.5%	7.9%

This is obviously a completely hypothetical and simplistic analysis. So far no-one has proposed that CCS is even capable of capturing all of the worlds CO_2e emissions.

However, it does illustrate a couple of points of interest:

- At a theoretical cost of only 1.9% of GDP, the West is reasonably positioned to be able to afford some level of carbon mitigation. But this cost rises to 6.1% of GDP for the Non-West.

- For India (the poorest of the major countries in the analysis) the cost would be around 8% of GDP.

The question is, will (or can) the West or Non-West commit such proportions of GDP to carbon mitigation?

CHAPTER 10

Politics of emissions

The debate over pricing carbon has been reignited after the Government's Environment spokesman confirmed – and then backtracked on comments - that introducing a carbon price for power companies would be considered as part of a climate change review.

It's a politically touchy topic — the ousting of Malcolm Turnbull as opposition leader in 2009 and Kevin Rudd as prime minister in 2010 can both be attributed in part to positions on emissions trading schemes.

http://www.abc.net.au/news/2016-12-05/the-carbon-pricing-debate-explained/8092506

In the introduction I said that Australia's current emissions policy crisis isn't just about emissions. Because fossil fuel exports are so important to the Australian economy, it's impossible to separate energy and emissions policy from economic policy. Energy policy needs to encompass not only emissions, but also the direction the entire economy should take over the next 25 years.

As I said in Chapter 2, while individuals can influence their own carbon footprint by changing their consumption habits, overall emission levels are determined by the economy as a whole. In other words, by political policy.

My aim with this book is to help you understand how carbon emissions work and their link to energy consumption. Specifically I wanted to provide insight into these areas of the emissions policy debate:

1. Has Australia done too much, the right amount, or too little to reduce greenhouse gas emissions over the past 28 years of the Kyoto Protocol?

2. Is renewable energy really cheaper than fossil fuel on a dispatchable or on-demand basis?

3. Should Australia be transitioning towards a zero-emission economy by the second half of this century?

4. Is it even possible for Australia to transition towards a zero-emission economy by the second half of this century?

5. Should Australia increase the price of petrol and diesel by around 70 c/L by imposing higher duties to the same level of taxation as in the UK and Western Europe?

6. Would such a proposal get through an election?

7. Aluminium smelting accounts for around 7% of Australia's emissions. Should aluminium production in Australia stop, either through a carbon reduction mandate or a carbon tax?

8. Should Australia stop exporting high direct carbon emission intense materials such as coal and liquefied natural gas (LNG)?

9. Should Australia stop exporting high indirect carbon emission intense materials such as iron ore and bauxite/aluminium ores?

10. Australia's contribution to worldwide greenhouse gas emissions is about 1.2%. Would any action taken by Australia to reduce CO_2e emissions have any significant impact on global emissions, either directly or by influence?

11. Some administrations (most notably the USA) have come to the conclusion that emission reductions can only be achieved at a cost to their economy they aren't willing to bear, and have pulled out of or are revising their emission reduction commitments. Should Australia follow suit?

Like most political issues, none of the answers to these questions are black and white. Would increasing petrol and diesel taxes to discourage air travel and ownership of high-consumption Sport Utility Vehicles (SUVs) harm Australia's economy? Probably not.

Would such a policy win an election? Again, probably not.

By now you've probably formed your own view on the issue. But one thing seems pretty clear. There's no 'silver bullet' solution for emissions reduction.

And what about the elephant in the room I mentioned in the introduction. As I explained in Chapter 7, over the past 30 years of the Kyoto Protocol, Australia has been expanding fossil fuel exports, with current volumes resulting in about 950 $MtCO_2$ per annum being emitted. Add in indirect high emission intense materials such iron ore and bauxite, and the emissions from Australia's exports dwarf our internal emissions.

Any emission reductions from reduced coal consumption in Australia will do nothing for global emissions if the coal displaced is simply exported to be burned elsewhere. The biggest impact Australia could have on global emissions is to wind down fossil fuel exports as quickly as possible. However, this would require a complete and radical transformation of the Australian economy.

Generally, countries that have reduced fossil fuel production have *not* done so for the reason of reducing carbon emissions. In the UK, recent fossil fuel production has reduced because the country has started to run out of cheap reserves. Even though the UK has one of the most ambitious emission reduction targets in the world, the UK Government's policy is to maximise recovery of its remaining oil and gas reserves. It's as if the UK is trying to squeeze out as much of the remaining fossil fuel revenue it can before fossil fuel usage is no longer considered acceptable. As far as I am aware, no country has yet taken the position of not producing fossil fuels that can be economically extracted. To propose this for Australia might be a political strategy with a significant degree of difficulty.

It seems obvious that globally emissions can't be addressed without tackling population and consumption. But it's unlikely that emissions can be reduced at no cost to the economy, either for individual countries or globally.

The lack of policy direction on energy can be traced back to one basic problem: what to do about carbon emissions. For the past 20 years, since the negotiation of the 'Australia Clause', successive Federal Governments have been unable to develop any form of coherent plan for what to do about Australia's emissions. Maybe the hope was the problem would somehow go away without having to take on the challenge. Maybe international climate efforts would all eventually break down and they wouldn't need to address the issue. With countries such as Canada, USA and Russia all at some stage pulling out of various commitments to take on emission reduction targets over the last eight years, it's still a possibility.

The debate about energy and emissions policy is essentially about two key issues. Unfortunately, they appear to be on a collision course.

1. As I said in Chapter 4, the way Australia negotiated its first rounds of emission targets meant it could easily meet the commitment without reducing consumption. There was no long-term requirement to progress tangible carbon emission reduction initiatives through either fuel efficiency improvements or a fuel mix change from coal to gas or greater levels of renewable energy.

   For the Australian coal industry, Kyoto Stage 1 meant business as usual. In fact, coal-fired generation increased over the period. Australia can't meet the far more onerous 27% Paris emission reduction target in this manner. Fossil fuel consumption reduction and generation fuel mix changes must be put in place.

2.   Some claim that renewable energy is already cheaper than fossil fuel. If that's the case, reducing emissions would be easy, and renewable energy should be booming without the need for government subsidy.

However, the official studies into the cost of electricity in Australia don't support this position. And while they forecast renewable energy being cheaper in the future (somewhere around 2030), they conclude that coal-fired power generation is currently the cheapest form of electricity. And that's without taking into account it isn't the average (or levelised) cost of electricity that is the key driver in the cost to the consumer. It is the marginal (or incremental) cost at peak demand.

This is the unstoppable force on a collision course with the immoveable object. It's the seemingly unsolvable problem that's been central to the demise of three Australian Prime Ministers (Kevin Rudd and Julia Gillard over emissions policy and the carbon tax, and Malcolm Turnbull over the 2018 NEG emissions policy).

The Australian political system and emissions policy

A major contributor to Australia's emissions policy indecision is the Australian political system itself. In fact, it may even be the root cause of the problem.

Like I said, a National Emissions Policy can't be just about emissions. It needs to cover emissions, energy *and* exports because they're all intimately connected. It is a broad-reaching brief.

The system of government in Australia is a two-tier system: Federal and State. Federal terms run up to three years, while State terms run for up to four years.

One thing hindering a coherent national energy and emissions policy is the need for a long-term view – 15 years at least, preferably 25. But with only three-year political terms at the Federal level, it's hard to see how this can be achieved. A national energy policy that favours coal for three years, renewables for three year and then coal again may not seem very coherent.

Another hindrance is how resources and production are controlled. According to Australia's constitution, control of natural resources (including fossil fuel extraction and use) falls under the jurisdiction of the States and Territories. They:

- Have final approval of resource extraction projects
- Collect royalty taxes on extraction of fuels and minerals
- Have control over power generation facilities and the fuel mix they use (by controlling fossil fuel extraction).

So the State Governments control Australia's energy use and resulting emissions, and the Federal Government is responsible for fulfilling Australia's emission reporting obligations to the IPCC. Apart from granting subsidies to renewable energy initiatives, the Federal Government can do little to influence emissions other than to set duties and taxes. And even that's a point of contention.

In late 2018, the price of petrol on the east coast of Australia spiked to about 160 c/L, primarily due to a short-term spike in the price of crude oil to about 75 US$/barrel. This reportedly led to a short-term reduction in private car use, and an increase in public transportation use. According to one media outlet, a State Government spokesperson laid part of the blame for the high petrol prices on the Federal Government's imposition of excise duties of about 50 c/L (including gst), and called for a reduction in the tax.

In the emissions reduction debate, it seems we want to have our cake and eat it too.

Another interesting example of State and Federal Governments lacking coherence and cooperation in energy and emissions policy is the Coal Seam Gas to Liquefied Natural Gas (CSG LNG) export projects brought online in Queensland in 2015.

Because each original Australian-owned proponent wanted to be the world's first CSG to LNG project, they wouldn't cooperate on a joint venture project. But the projects were so big ($25 billion each) that they could only be financed by selling around 85% of the projects and associated gas reserves to overseas national and multinational oil companies. None of this seemed to raise alarm bells at either a state or federal level. To be fair, overseas ownership of gas reserves may not be a concern providing the internal supply market still operates reasonably efficiently.

However, their desire to have their own projects also led to them installing too much capacity – two LNG facilities each (called "LNG trains") for a total of six, instead of three or four shared LNG trains – creating more LNG export capacity than the gas fields can comfortably supply. With export LNG attracting a sales price of about 12-18 $/GJ (it's linked to crude oil, so it varies), and excess capacity in the export supply chain, it has effectively 'sucked' any excess domestic gas into the export market and driven up the price for domestic users across eastern Australia from about 4 $/GJ to parity with export prices. This adds about 5 c/kWh to the price of electricity from natural gas, making a transition from higher-intensity coal to natural gas more expensive than it would have been under pre-export pricing.

This, in combination with the NSW and Victorian State Governments imposing severe restrictions on onshore gas exploration, pretty much caused the gas supply crisis in Australia in 2017. It seemed to occur without any Federal or State Government oversight, and without considering how the availability of a lower-emission-intensity fuel would affect national emission targets. There was no real energy or emission policy in place to test approval of the gas export projects against.

In 2017 the Federal Government tried to intervene in the gas crisis, calling the national and multinational oil companies into negotiations and threatening to restrict exports if gas supplies weren't made available to domestic users. But supply isn't really the issue here – there's plenty of gas available. The issue is price. The gas is going to the highest bidder: the export market.

> "*Gutless, arrogant and disrespectful. Worse than Tony Abbott. He announces a gas plan when he hasn't had the common decency to talk to me when he wants to steal our (gas) royalties.*"
>
> **As reported in Brisbane Sunday Mail 30/4/17.**

The quote is attributed to the Queensland Premier Annastacia Palaszczuk, and is referring to Prime Minister Malcolm Turnbull following the Federal Government's attempted intervention in the 2017 gas supply crisis.

A political system where the states have primary control over fossil fuel extraction and power generation policy has resulted in these contradictory energy policy outcomes across eastern Australia:

- A former South Australian Labor Government chased ever-increasing wind power generation, then needing to build new fossil gas and diesel generation capacity to back it up *and* encouraging increased on-shore gas exploration in the state.

- A Victorian Labor Government imposing severe restrictions on on-shore gas exploration activity.

- A New South Wales Liberal Government maintaining restrictions on on-shore gas exploration activity, and at the same time considered liquefied natural gas (LNG) import facilities to import what was likely to be imported fracture stimulation produced gas. (Fracture stimulation (fraccing) is currently restricted in NSW and Victoria.)

- A Queensland Labor Government promoting large-scale solar PV generation while at the same time promoting a major new fossil fuel export project (the 40 Mt/year Adani coal mine) on the grounds of jobs and economic growth.

- The Federal Government's role in all of this is really only to negotiate targets with the international community and to calculate and report the actual greenhouse gas emissions. The only control it can exercise over emissions and energy policy is to set fuel excise duties (and a possible carbon tax, if that were to be back on the agenda), and subsidise renewable energy.

I doubt anyone could devise a less coherent supporting political structure to a national energy and emissions strategy.

To quote again from the Independent Review into the Future Security of the National Electricity Market[1]:

> The uncertain and changing direction of emissions reduction policy for the electricity sector has compromised the investment environment in the NEM (National Electricity Market). The lack of a transparent, credible and enduring emissions reduction mechanism for the electricity sector is now the key threat to system reliability. Without investment in new generation capacity, the reliability of the NEM will be compromised. It is critically important that there is widespread political and community acceptance of the need for a stable policy framework.

If Australia's political system doesn't seem amenable to developing stable and coherent national energy and emission policy, what needs to be put in place to enable one?

It seems we might be reduced to two options:

1. Greater Federal Government control over the States' natural resources management.

2. Less Federal Government control over the States' natural resources management.

State control of natural resources is enshrined in Australia's constitution. It's hard to imagine them ever relinquishing that control, and so perhaps the only viable solution is to let the states go it alone and have the Federal Government relinquish energy and emissions policy management altogether.

If this was done, it would also seem appropriate for the states to be accountable for their own emissions reporting and have their own targets within the international community.

A price on carbon?

When the Federal Labor Government introduced a carbon tax in 2011 (at only 23 $/tCO$_2$e), it was delivered with the political message that industry should not pass the costs onto the consumer. That seems like raising the duty on cigarettes to deter smoking, but expecting the tobacco companies to not pass the costs onto the end user. The fact is when a society wants to deter consumption of a product, it imposes a tax on it.

The 2011 carbon tax was applied only to electricity and industrial greenhouse gas emissions. Petroleum liquid transportation fuels and all other emission sectors were exempt, including agriculture. Why should electricity consumers have to pay a carbon tax when consumers of high-emission motor vehicles and air travel were exempt? As most consumers of these emission sources would tend to be at the wealthier end of the economy, this seems like a tax concession for the well off.

Successive Federal Governments approach to emission reductions have generally been promoting renewables into the electricity market through subsidies. As I discussed in Chapter 5, this approach is throwing a spanner into the works of how free market competition is supposed to bring cost reduction benefits to the consumer. It's almost as if the Government has put itself into a position of competing with the fossil fuel industry.

In my view, the fairest and most transparent way to incentivise reductions in greenhouse gas emissions would be an across-the-board carbon price, or tax. It should be applied across all emission sources, including transportation fuels (petrol, diesel and air travel) and agricultural production. The effect of an example 100 $/tCO₂e carbon tax on the cost of energy products was shown at the end of Chapter 2.

The big problem of a unilaterally applied carbon tax is one of fairness for Australian manufacturers competing with overseas competitors that do not have a tax. To ensure Australian export industries aren't unfairly disadvantaged by such a tax, all activity associated with producing export goods could be exempt. This would include emissions associated with extraction of natural resources and fossil fuels for export. (Chapter 7 showed how Australian exports contribute significantly to global emissions, but it would be up to other jurisdictions to determine whether they impose their own carbon tax on Australian fossil fuel imports. This of course would be a significant point of debate).

To ensure Australian industries producing for the in-land market aren't unfairly disadvantaged, all imports would need to be taxed with an equivalent carbon import tax based on the carbon footprint needed to produce the goods.

While the overall cost of everything that has an emission footprint (including imported goods) would increase, Australian exporters and in-land manufacturers would not be disadvantaged. This approach would avoid driving carbon emissions to be outsourced to other jurisdictions.

The real advantage of a carbon tax is it gives transparency to the electorate of the cost of each political party's emission reduction policies.

The problem with this approach is it's very impractical to accurately determine the carbon footprint of specific wholesale imported goods. While the carbon emissions for whole economies can be estimated with reasonable accuracy, allocating them to specific units of production would require immensely complicated allocation rules and material traceability requirements. And the possibility that such a scheme could be worked into a global free trade framework seems unlikely.

Summary

It is not possible to to develop a policy for emissions without a long term plan for energy consumption. As I've shown in this book, it is relatively simple to link most of our emissions with the associated consumption of fossil fuels.

Reporting of fossil fuel production and consumption statistics directly along-side international emission reporting obligations would greatly improve the transparency of emissions performance reporting. Not just for Australia, but for all countries. It would provide a base line for a realistic and coherent plan for energy use and emissions, whether it be targeting reductions, or maintaining the status quo.

My view is we will need some form of commission or inquiry to find a way past the issues that will otherwise keep hindering a national emissions and related energy policy and plan. As I've discussed, it also needs to include consideration of fossil fuel extraction and exports. There is no point in Australia reducing fossil fuel emissions if the fuel displaced is simply exported to be burnt elsewhere. The problem is such an inquiry may reach a similar conclusion to the US Administration. Trying to make carbon emission reductions that are meaningful in a global context may present significant risks to the economy. The question is whether those risks are worth taking, given that the chance of mitigating carbon emissions on a global scale seems quite small.

CONCLUSION

While there are still many climate sceptics, there appears to be a consensus in the scientific community that global temperatures are changing – and in the opposite direction to that expected by historical Earth system cycles, which indicate the Earth should be cooling.

While the ever-changing geology of the Earth (plate tectonics and volcanic activity) makes it extremely difficult to make conclusive measurements, there also appears to be a consensus that sea levels are rising. Estimates are that they have risen by about eight inches (200mm) in the past 100 years. And current global warming scenarios indicate they could rise by 0.5-2m over the next 100 years.

In debating humanity's response to global warming, we need to consider two big issues:

- Can anything actually be done about emissions on a global scale?
- How serious is this on an evolutionary scale?

I've covered the first issue in this book. Now let's consider the second issue.

Around only 18,000 years ago, the Earth's surface that now forms the Great Barrier Reef was on dry land. The sea level at that time – the last glacial maximum (or Great Ice Age) – is estimated to be 120m below current levels. Homo sapien populations were in decline in Northern Europe and Asia due to thick layers of glacial ice covering much of the land. But the low sea level allowed a second wave of homo sapiens to colonise Australia due to land bridges and narrow sea crossings from Southeast Asia.

In nature, every challenge usually presents a corresponding opportunity.

An estimated 99.9% of the species that have existed on Earth are now extinct. In many cases these species evolved to successfully exploit a brief environmental niche or opportunity. Human's exploitation of fossil fuels may eventually belong in this category. Nature has a habit of not letting success proliferate, and the history of evolution demonstrates a propensity for replacing previous top dogs. Do homo sapiens have any more right to prolonged existence on the planet that any other form of life on Earth?

Despite energy and resources companies producing sustainability reports as part of their modern new-age reporting obligations, there's actually no such thing as sustainable consumption of fossil fuel. Fossil fuels accumulated over millions of years, and we are consuming them over only a few hundred. Human existence on Earth can only be considered truly sustainable when it can survive on only 100% renewable energy. This may need to include traditional sources of renewable energy, such as horse power and sustainably grown bio-combustion fuel and building material (sustainable timber).

It's hard to believe we'll be able to convert to low-emission technology in some sectors of energy consumption, and still have access to energy use privileges only fossil fuel can provide such as global air travel.

It seems unlikely that we could find a way to reduce either short-term population growth or global fossil fuel consumption and emissions. We're heading towards a future world that may well be a less comfortable place for many humans to exist. This may ultimately put a natural limit on the population the planet can sustain, and the homo sapien population may need to return to a level truly sustainable consumption can support. That may be less than two billion, or even less than one billion.

In developing advanced technologies, primarily through exploiting fossil fuel energy, homo sapiens might be considered the first species that could opt out of the natural selection process. Global warming may just be Nature's way of opting us back in.

REFERENCES

1. Independent Review into the Future Security of the National Electricity Market: Blueprint for the Future, Commonwealth of Australia 2017.

2. Mike Burners-Lee and Duncan Clark, The Burning Question. Published by Profile Books, 2013

3. National Inventory Report 2015 Volume 1, Commonwealth of Australia 2017.

4. National Inventory Report 2016 Volume 1, Commonwealth of Australia 2018.

5. Quarterly Update of Australia's National Greenhouse Gas Inventory: June 2016, Commonwealth of Australia 2016.

6. Quarterly Update of Australia's National Greenhouse Gas Inventory: June 2017, Commonwealth of Australia 2017.

7. National Inventory Report 2014 (revised) Volume 1, Commonwealth of Australia 2016.

8. Australian Petroleum Statistics, Commonwealth of Australia 2017.

9. Australian Energy Update 2016, Department of Industry, Innovation and Science, Canberra. (The Commonwealth of Australia does not necessarily endorse the content of this publication).

10. Australian Land Use, Land Use-Change and Forestry emissions projections, Commonwealth of Australia March 2015.

11. 2015 UK Greenhouse Gas Emissions, Final Figures. Crown copyright 2017. Used under the terms of the Open Government Licence.

12. Digest of United Kingdom Energy Statistics (DUKES) 2016. Crown copyright 2016. Used under the terms of the Open Government Licence.

13. Australian Power Generation Technology Report 2016. Produced by a consortium led by CO2CRC Consultants, with Steering Committee of CO2CRC, CSIRO, Australian Renewable Energy Agency, Department of Industry and Science – Office of the Chief Economist, Anlec R&D.

14. Pollution and Price: The cost of investing in gas. By Andrew Stock, Professor Will Steffen, Greg Bourne, Petra Stock and Dr Martin Rice. The Climate Council of Australia Ltd.

15. Resources and Energy Quarterly September 2016 (Vol. 5, no.5). Commonwealth of Australia 2016.

16. Energy Communications Guide 2018. Authors: Alexia Boland, Petra Stock and Louis Brailsford. The Climate Council of Australia Limited.

17. Arif Syed (BREE). The Australian Energy Assessment (AETA) 2013 Model update. Canberra, December 2013. Copyright Commonwealth of Australia 2013.

18. Australia's emissions projections 2017, Commonwealth of Australia 2017.

19. Nasa: https://earthobservatory.nasa.gov/Features/CarbonCycle

20. http://coal.makingthefuturepossible

21. http://www.environment.gov.au/climate-change/publications/factsheet-australias-2030-climate-change-target

22. Yoichi Kaya, Keiichi Yokobori. Environment, Energy, and Economy: strategies for sustainability. as the output of the Conference on Global Environment, Energy, and Economic Development (1993: Tokyo, Japan).

23. epa.vic.gov.au/agc/calculator/index.html

24. Wikipedia contributors. List of countries by past and future population. In Wikipedia, The Free Encyclopedia. Citing: United States Census Bureau - International Data Base (IDB), July 2015 edition.

25. Wikipedia contributors. List of countries by greenhouse gas emissions. In Wikipedia, The Free Encyclopedia. Citing: "Climate Analysis Indicators Tool (CAIT) Version 2.0. (Washington, DC: World Resources Institute, 2014)". World Resources Institute.

26. Brown, T. J.; Idoine, N. E.; Raycraft, E. R.; Shaw, R. A.; Hobbs, S. F.; Everett, P.; Deady, E. A. and Bide, T. 2018. World Mineral Production: 2012–2016. British Geological Survey, Keyworth, Nottingham. ISBN 978-0-85272-882-6

27. Wikipedia contributors. List of countries by steel production. In Wikipedia, The Free Encyclopedia. Citing: "World Steel Production 1980-2013". World Steel Association, Nov 2014; "Steel Statistical Yearbook 2016". World Steel Association, 2016.

28. Wikipedia contributors. List of countries by GDP (nominal). In Wikipedia, The Free Encyclopedia. Citing: "World Economic Outlook Database". International Monetary Fund. April 2018.

29. Gold Standard Validation Report. Project title: SSEE1 Solar PV 1 – 10 Power Plant Project, Thailand. Report no CCMS/35609. Report version no. 03, 2017. By URS Verification Private Limited, Noida, India.

30. Australian Energy Market Operator (AEMO). https://www.aemo. com.au/Electricity/National-Electricity-Market-NEM/Data-dashboard#aggregated-data/PRICE_AND_DEMAND_201801_ QLD1

31. https://www.edie.net/news/10/Biomass--carbon-neutrality--debate-continues-to-divide-opinions/

32. Australian Aluminium Council. aluminium.org.au/climate-change/ aluminium-smelting-greenhouse-performance/

33. http://www.climatechangenews.com/2016/05/31/russia-moots-change-to-2030-emissions-target/. https://www.carbonbrief.org/ ambiguous-russian-climate-pledge-mystifies-many

34. https://www.solarchoice.net.au/blog/solar-calculator-resource-library

35. NEM outlook and Snowy 2.0. Report prepared for Snowy Hydro Limited. A Marsden Jacob Final Report.

36. Howarth, Robert W. A Bridge To Nowhere: Methane Emissions and the Greenhouse Gas Footprint of Natural Gas. Department of Ecology & Evolutionary Biology, Cornell University, Ithaca, New York, 15 May 2014.

APPENDIX A

Domestic solar PV and
battery calculations

In Chapter 5 we saw this chart showing the typical generation characteristic for a domestic solar photovoltaic (PV) power system overlayed on a typical household usage pattern.

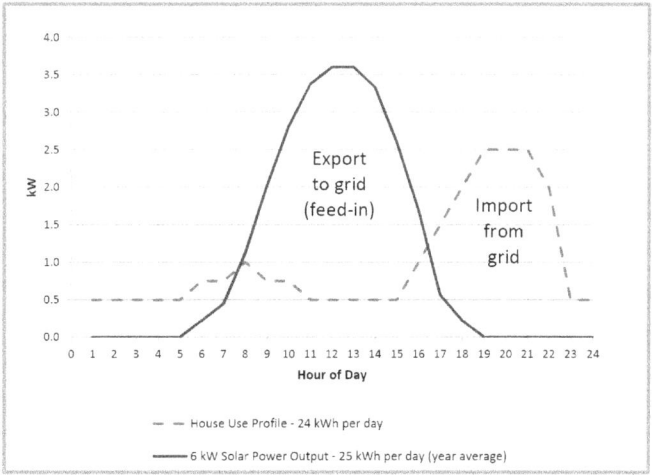

Source [34]

The solid line shows the average solar PV power generation output for a 6 kW nameplate capacity system. The curve is the average output over an entire year for a north-facing installation in Brisbane, Queensland.

The vertical axis shows the power (in kW) at each hour of the day along the horizontal axis, and so the area under each curve equates to the kilowatt-hours being produced or consumed.

The total average energy output for the solar PV system is 25.1 kWh per day—an average power output of 1.046 kW. In Chapter 5, we defined the capacity factor as the average power output divided by the nameplate capacity. In this example the average capacity factor (CF) is 1.046 / 6.0 = 0.174, or 17.4%.

The dashed line shows the household consumption pattern. The area under this curve is 24 kWh per day, which is an average power requirement of 1 kW. So at 1.046 kW, the 6 kW solar PV system is generating slightly more power than the household is consuming.

However, the household usage is not a constant power requirement. Overnight it's about 0.5 kW, rising to about 1 kW as the household wakes up to its morning routine. Daytime consumption drops back down to about 0.5 kW. But the evening usage starts to rise about 4pm, peaks at 2.5 kW between 7pm and 9pm, and then falls back to the overnight usage of 0.5 kW.

As you can see, most of the solar PV energy generated isn't actually used by the household. Only 7.2 kWh of the solar power is used directly by the house (around 30%.) The remaining 17.9 kWh is exported back into the local power grid (as labelled in the chart). This is termed 'Feed-in' power, and the price the household gets for this electricity is called the 'Feed-in tariff' (FIT).

So how much money is this solar PV system actually making? We need to consider two things:

- The solar power being used directly is saving power that otherwise would need to be bought at the normal retail price of electricity.
- The solar power being exported is generating revenue at the agreed feed-in tariff price.

As most of the power in this example is 'Fed-in' (70%), the feed-in tariff is the major factor in the economics of the solar PV system.

The peak electricity supply tariff in Queensland is currently 25 c/kWh excluding GST. However, most retailers now offer pay-on-time discounts of around 28%. So the retail price of electricity is:

25 x (1 – 0.28) x 1.1 = 19.8 c/kWh. (To make the calculations easier we'll round it up to 20 c/kWh.)

Feed-in tariffs tend to be more variable. In the early days of solar, FITs were heavily subsidised by governments to encourage take up. Some were as high as 50 c/kWh—more than double the retail price. Now they tend to be 6–12 c/kWh. (Here we'll use 10 c/kWh, which is 50% of the retail price.)

So at 20 c/kWh retail electricity price and 10 c/kWh FIT, the 6 kW solar PV installation is making (on average):

> 7.2 kWh/day x 0.20 \$/kWh = 1.44 \$/day on consumed solar
>
> 17.9kWh/day x 0.10 \$/kWh = 1.79 \$/day on exported solar
>
> Total   = 3.23 \$/day

If the cost of the solar PV system is \$6,000 (\$1,000 per kW), this installation has an average payback time of:

> \$6,000 / (3.23 x 365) =   5.1 years straight payback (excluding the effect of interest rates).

This might be considered a reasonable investment. After five years the installation pays for itself and generates net income for the household. But how much income it will generate over the next 15 years or so depends on the future feed-in tariff price. The economics of standalone solar depend on being able to sell the peak power generated by to a retailer, which then sells it on to another consumer. If the market becomes too saturated, the demand for feed-in power may reduce with a corresponding reduction in the value of the feed-in tariff.

To accurately determine the benefits of solar, the consumer must estimate how much of their daily solar generations can be used directly rather than exported at a lower feed-in tariff price.

Domestic Battery Installations

In November 2018, the Federal Labor Party announced a new energy and emissions policy. A significant component of the policy was to provide subsidies of up to $2,000 to households for installing domestic batteries to augment solar PV systems.

These domestic batteries store part or all of the solar PV energy that would otherwise be exported. Using the previous 6 kW solar PV system as an example, and assuming that all of the previously exported solar energy is stored, 17.9 kWh of storage would be needed.

Unfortunately, batteries are not 100% efficient. On average they return about 90% of the energy put into them. So the actual battery would need to be about 19.9 kWh (17.9/0.9). To cover this inefficiency, the solar PV system would also need to be about 10% larger than the original 6 kW system (based on average yearly production).

In winter, capacity factors fall to an average of 13.6. This means a larger battery system may be needed if the household's electricity requirements don't reduce by the same amount. This doesn't allow for storage during any periods of below average production, and so this demand would still need to be met by importing electricity from the grid.

So how much money is the new combined solar PV and battery system making now?

Assuming on average the household's power requirement can now be fully met by the solar/battery combination; there will be no need to import power from the grid. So the house will save all its retail electricity cost:

24 kWh/day x 0.20 $/kWh = 4.80 $/day

While they are forecast to reduce, the cost of domestic solar storage batteries is currently around $1,000 per kWh. So the total costs for the combined solar/battery installation will be:

6.6 kW x 1,000　　= 　$6,600 for the larger solar PV system

19.9 kWh x 1,000 = 　$19,900 for the 90% efficient battery

Total　　　　　　= 　$26,500 for the combined system

The combined solar/battery installation has an average payback time of

$26,500/ (4.8 x 365) = 　15.1 years (straight payback, excluding the effect of interest rates)

As you can see, the payback on the combined installation is significantly worse than the five-year payback on the standalone solar PV installation. For now we're ignoring the effect of the proposed $2,000 policy subsidy.

But this figure is the payback on the combined solar PV/battery installation. To estimate the true payback on the battery, we must consider the marginal (or incremental) cost and benefit of just the battery installation decision over the previous standalone solar PV installation.

The marginal cost of just the battery installation is:

$26,500 - $6,000　 = 　$20,500

The marginal saving or benefit from just the battery installation the benefit (saving) of the battery/solar combined system minus the benefit of the solar PV standalone installation.

$4.80 - $3.23 　　= 　$1.57 per day

Which means the battery only installation has an average payback of

$20,500/ (1.57 x 365) = 　35.8 years (straight payback, excluding the effect of interest rates)

Even with a government subsidy of $2,000 per $10,000 of battery costs (20%), the marginal payback on a battery installation would be improved by 20% to 28.6 years.

While a payback of more than 20 years for the marginal battery installation may not be considered a worthwhile investment, a consumer might accept a low-return investment for the sake of emissions reductions. The problem is the batteries are currently sold with a guaranteed lifetime of only nine or ten years. If the battery needs to be replaced before it achieves its marginal payback, the consumer won't just fail to recover the costs. They'll actually lose money.

Battery marginal installations can provide such low rates of return because they're sensitive to the value of the feed-in tariff. For standalone solar, the higher the feed-in tariff the better the return and payback.

For example, if the feed-in tariff was set at the retail price of electricity, the solar PV system would cover the entire cost of electricity for the house without needing a battery (even though it still wouldn't produce all the electricity the house uses). The payback would improve to 3.3 years.

But with a battery the opposite is true. The higher the feed-in tariff, the less benefit (saving) the battery provides. There's no payback at all (technically it's infinity) on batteries with a feed-in tariff set at the retail price of electricity.

A zero feed-in tariff would improve the marginal payback on the incremental battery installation to 13 years, while the payback on the standalone solar PV would deteriorate from 5.1 years to 11.4 years.

Predicting the actual feed-in tariffs over the payback period required for a battery might be considered difficult. One power retailer in Australia is currently offering a 20 c/kWh feed-in tariff for two years. As we saw in the previous section, this makes the payback on standalone solar very attractive. However, it eliminates the incentive to install battery storage over the two years of the offer.

It's another example of why implementing increased volumes of renewable energy with storage needs to be coordinated under a combined energy and emissions policy.

APPENDIX B

Levelised cost of Electricity in the *Independent Review into the Future Security of the National Electricity Market: Blueprint for the Future*, Commonwealth of Australia 2017[1]

The following chart is reproduced from data contained in the chart on page 201 of the report.

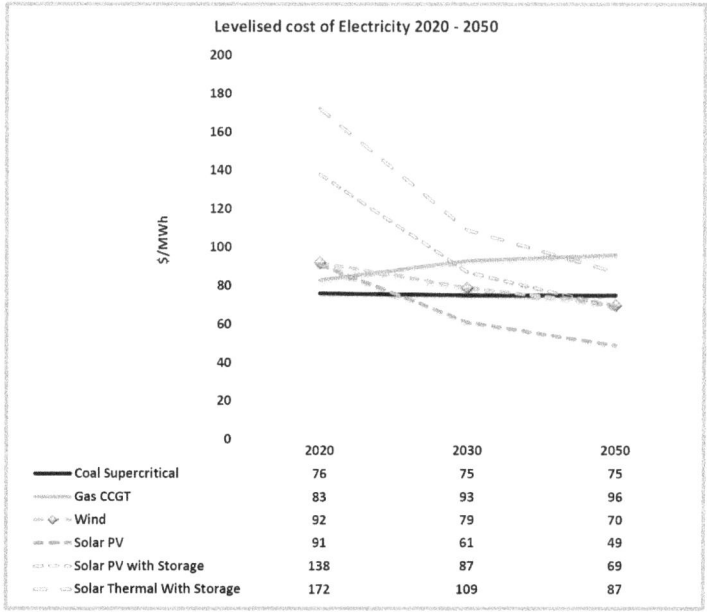

Levelised cost of Electricity 2020 - 2050	2020	2030	2050
Coal Supercritical	76	75	75
Gas CCGT	83	93	96
Wind	92	79	70
Solar PV	91	61	49
Solar PV with Storage	138	87	69
Solar Thermal With Storage	172	109	87

The three columns for each technology type show the costs assumed in 2020, 2030 and 2050.

Notes (from the report):

- The numbers in Figure A.1 refer to the average.

- Large-scale Solar Photovoltaic includes fixed plate, single and double axis tracking.

- Large-scale Solar Photovoltaic with storage includes three hours' storage at 100% capacity.

- Solar Thermal with storage includes 12 hours storage at 100% capacity.

- Cost of capital assumptions are consistent with those used in policy cases (i.e. without the risk premium applied).

- The assumptions for the electricity modelling were finalised in February 2017 and don't take into account recent reductions in technology costs (e.g. recent wind farm announcements).

This shows that while the report indicates that solar PV and wind power generation are forecast to be cheaper by 2030 and 2050, black coal fired power generation will be the cheapest from of electricity until into at least the 2020s.

ABOUT THE AUTHOR

Andrew Perry graduated from the University of Bradford with a degree in Chemical Engineering. He has worked for 30 years as a technical specialist in the energy and resources industries. His career started in the UK in high-conversion oil refining, specialising in energy efficiency and conversion. In Australia he has worked in oil and gas exploration and production, including on Queensland's Coal Seam Gas to Liquefied Natural Gas export projects. His roles have included design and construction, project management, and operations optimisation. He has previously published technical papers on optimising oil refinery operations and engineering design.

He works as an independent consultant in the natural resource and energy sector.

CPSIA information can be obtained
at www.ICGtesting.com
Printed in the USA
BVHW041116310119
539142BV00015B/1587/P

9 780987 635808